Emerging Technologies and Digital Transformation in the Manufacturing Industry

Fernando Ortiz-Rodriguez
Tamaulipas Autonomous University, Mexico

Sanju Tiwari
Tamaulipas Autonomous University, Mexico

Luis Manuel Hernández-González
INDEX, Mexico

Agustin Tiburcio
INDEX, Mexico

A volume in the Advances in Logistics, Operations, and Management Science (ALOMS) Book Series

IGI Global
PUBLISHER of TIMELY KNOWLEDGE

Published in the United States of America by
IGI Global
Business Science Reference (an imprint of IGI Global)
701 E. Chocolate Avenue
Hershey PA, USA 17033
Tel: 717-533-8845
Fax: 717-533-8661
E-mail: cust@igi-global.com
Web site: http://www.igi-global.com

Library of Congress Cataloging-in-Publication Data

Names: Ortiz-Rodriguez, Fernando, 1974- editor. | Tiwari, Sanju, 1979- editor. | Hernandez-Gonzalez, Luis Manuel, 1973- editor.
Title: Emerging technologies and digital transformation in the manufacturing industry / edited by Fernando Ortiz-Rodriguez, Sanju Tiwari, Luis Manuel Hernandez-Gonzalez, Agustin Tiburcio-Sanchez.
Description: Hershey, PA : Business Science Reference, [2023] | Includes bibliographical references and index. | Summary: "This book deals with the manufacturing process using a technology-driven approach that utilizes Internet-connected machinery to monitor the production process, artificial intelligence, sensors, and emerging technologies that lead to efficiency"-- Provided by publisher.
Identifiers: LCCN 2023034546 (print) | LCCN 2023034547 (ebook) | ISBN 9781668480885 (hardcover) | ISBN 9781668480892 (paperback) | ISBN 9781668480908 (ebook)
Subjects: LCSH: Manufacturing industries. | Manufacturing processes--Technological innovations/ | Information technology--Management. | Organizational effectiveness.
Classification: LCC HD9720.5 .E524 2023 (print) | LCC HD9720.5 (ebook) | DDC 338/.064--dc23/eng/20230724
LC record available at https://lccn.loc.gov/2023034546
LC ebook record available at https://lccn.loc.gov/2023034547

This book is published in the IGI Global book series Advances in Logistics, Operations, and Management Science (ALOMS) (ISSN: 2327-350X; eISSN: 2327-3518)

British Cataloguing in Publication Data
A Cataloguing in Publication record for this book is available from the British Library.

Advances in Logistics, Operations, and Management Science (ALOMS) Book Series

John Wang
Montclair State University, USA

ISSN:2327-350X
EISSN:2327-3518

MISSION

Operations research and management science continue to influence business processes, administration, and management information systems, particularly in covering the application methods for decision-making processes. New case studies and applications on management science, operations management, social sciences, and other behavioral sciences have been incorporated into business and organizations real-world objectives.

The **Advances in Logistics, Operations, and Management Science** (ALOMS) Book Series provides a collection of reference publications on the current trends, applications, theories, and practices in the management science field. Providing relevant and current research, this series and its individual publications would be useful for academics, researchers, scholars, and practitioners interested in improving decision making models and business functions.

COVERAGE

- Marketing engineering
- Production Management
- Services management
- Finance
- Computing and information technologies
- Information Management
- Operations Management
- Organizational Behavior
- Risk management
- Political Science

The Advances in Logistics, Operations, and Management Science (ALOMS) Book Series (ISSN 2327-350X) is published by IGI Global, 701 E. Chocolate Avenue, Hershey, PA 17033-1240, USA, www.igi-global.com. This series is composed of titles available for purchase individually; each title is edited to be contextually exclusive from any other title within the series. For pricing and ordering information please visit http://www.igi-global.com/book-series/advances-logistics-operations-management-science/37170. Postmaster: Send all address changes to above address.

Titles in this Series

For a list of additional titles in this series, please visit: http://www.igi-global.com/book-series/

New Perspectives and Possibilities in Strategic Management in the 21st Century Between Tradition and Modernity
Javier Martínez-Falcó (University of Alicante, Spain & Stellenbosch University, South Africa) Bartolomé Marco-Lajara (University of Alicante, Spain) Eduardo Sánchez-García (University of Alicante, Spain) and Luis A. Millan-Tudela (University of Alicante, Spain)
Business Science Reference • © 2023 • 514pp • H/C (ISBN: 9781668492611) • US $315.00

Government Impact on Sustainable and Responsible Supply Chain Management
Atour Taghipour (Normandy University, France)
Information Science Reference • © 2023 • 354pp • H/C (ISBN: 9781668490624) • US $240.00

Transport and Logistics Planning and Optimization
Jaouad Boukachour (University of Le Havre, France) and Abdelhamid Benaini (University of Le Havre, France)
Business Science Reference • © 2023 • 317pp • H/C (ISBN: 9781668484746) • US $250.00

Black Women's Formal and Informal Ways of Leadership Actualizing the Vision of a More Equitable Workplace
Audrey D. Bowser (Arkansas State University, USA) Kimberley M. Davis (University of Arkansas at Pine Bluff, USA) and Natalie A. Johnson-Leslie (Arkansas State University, USA)
Information Science Reference • © 2023 • 300pp • H/C (ISBN: 9781668438275) • US $215.00

Considerations and Characteristics of the Millennial Leader in Today's Global Society
Tymon M. Graham (TMGraham Consulting, LLC, USA)
Business Science Reference • © 2023 • 300pp • H/C (ISBN: 9781668447116) • US $240.00

The Economics of Talent Management and Human Capital
Sharif Uddin Ahmed Rana (Paris Graduate School and World Talent Economy Forum, Malaysia) and Adrian David Cheok (Imagineering Institute, Malaysia)
Business Science Reference • © 2023 • 300pp • H/C (ISBN: 9781668466414) • US $235.00

IGI Global
PUBLISHER of TIMELY KNOWLEDGE

701 East Chocolate Avenue, Hershey, PA 17033, USA
Tel: 717-533-8845 x100 • Fax: 717-533-8661
E-Mail: cust@igi-global.com • www.igi-global.com

Table of Contents

Detailed Table of Contents

All of the advances that we see in our lives are mostly based on software. Over the last several years, software development techniques have seen a significant transition. It simplifies and profits the firm, for example. Many customers' burdens are lightened as a result of their combined efforts. Today, software powers practically every area of the economy and is an integral aspect of a country's economy. Artificial intelligence is no exception. Businesses have embraced practically all developing technology in their software development. AI's effect on software development alters how businesses operate and makes software smarter. In this chapter, the authors have talked about the change AI would bring to software development and the role of could. The authors have also discussed traditional ways used. Towards the end of this paper, they have added open issues and challenges in detail.

The technological advances that the world has experienced throughout all the years have caused important changes in the structures of society. Education is one of the fields in which the industrial revolution, driven by technological development, caused a significant impact, since technology allowed students to freely access more information and therefore more knowledge. Education 2.0 and 3.0 brought not only access to information, but also greater connectivity and interaction between people, particularly between teachers and their students. The dynamism of technology also brought opportunities for innovation with Education 4.0, changing the existing paradigm from face-to-face to online classes, the emergence of digital, your relationship with those around you, and your abilities to meet basic demands and roles in society. These changes meant the application of new educational methodologies, aimed at significant learning that prepares the individual with the necessary skills to face the modern world of work.

In today's global economy, emerging economies play a very important role in the production chain. Countries such as China and India have become competitive participants in the global market. The labor force in these emerging markets has developed new skills, leading developed economies such as the United States to move important areas of their production chain into these countries. This phenomenon had reached Mexico as well and multiple cases of salary and wage inequality could be found. Thus, this research tries to establish a frame that could help better understand the effect that salary and wage inequality play in the leadership-productivity relation among middle management supervisors within the manufacturing industry at the Mexico-US border. A questionnaire was developed and applied to middle management supervisors on both sides of the border (Mexico and the US). The main contribution of this study is to help better understand the impact of salary and wage inequality among middle management supervisors.

Emerging technologies, understood as new technologies or the continuous development of existing technologies, have led to significant advances ranging from the simplification of everyday tasks to the automation and optimization of industrial processes; and therefore, to increase the productivity and quality of life of people. Who does not like to have more time to devote to other activities or to rest, or to have some device that allows one to simplify life with a click. Or in the case of companies, reduce production times, costs, and process failures. While we may think that responsibility for promoting the progress of emerging technologies; as well as the definition of public policies, regulations, and regulation of this matter lies in the figure of the government, the reality is that the work and interference of other actors such as the private sector is crucial to its progress.

Preface

Manufacturers must adapt to secure customer loyalty driven by the changing expectations of new generations. Manufacturers globally were making some progress in transforming before the pandemic digitally. Pilots and scaling of emerging technologies like robots and machine learning (ML) were underway to build smarter, increasingly automated, and more cost-effective factories.

Digital transformations are different from regular business transformations, in both small and big ways. For one thing, business transformations usually end once a new behavior has been achieved. Digital transformations, on the other hand, are long-term efforts to rewire how an organization continuously improves and changes. That's because technology is not only becoming further integrated into business but also constantly evolving. For example, given the growing importance of AI in generating business insights and enabling decision-making logic, any digital transformation should also be an AI transformation.

Emerging technology is used to describe a new technology, but it may also refer to the continuing development of existing technology. The term commonly refers to technologies that are currently developing, or that are expected to be available within the next five to ten years, and is usually reserved for technologies that are creating, or are expected to create, significant social or economic effects. Emerging digital technologies have generated new opportunities while creating new legal challenges, particularly related to copyrights, trademarks, patents, royalties, and licensing.

Artificial intelligence is shaping the future of humanity and is starting to have an impact on the manufacturing sector. It is already the main driver of emerging technologies like big data, robotics, and IoT, generative AI, and it will continue to act as a technological innovator for the foreseeable future., forty-four percent of companies are looking to make serious investments in AI and integrate it into their businesses.

The book is organized into 14 chapters, aimed to draw part of the actual state of digital transformation and emerging technology in the manufacturing sector.

Chapter 1 deals with technology related to investment and finance. The world of manufacturing has shaped several countries, and regions and created specific sites of excellence in several places, however, the speed at which investment is being made requires that all support areas be agile and effective on the deliverables; finance plays a significant role from day-to-day activities to strategy, investment, and recently and after the covid 19 pandemic a key role in the relocation of manufacturing footprint. The speed of cost, rate of revenue, and inventory have become major challenges for companies and finance executives.

Chapter 2 aims to explain the efforts and needs to develop a software update, in a company's import or export transactions, you should know that there is a big world to know behind these and understand the inferences that must suffer by not having the proper administration through Customs software. Being a company of the IMMEX program in Mexico is critical and when you need to change software you will face a series of challenges with which will have to demonstrate the administrative skills of the Project Leader and the rest of the personnel involved.

Chapter 3 deals with emerging technologies, all the advances that we see in our lives are mostly based on software. Over the last several years, software development techniques have seen a significant transition. It simplifies, simplifies, and profits the firm, for example. Many customers' burdens are lightened because of their combined efforts. Today, software powers practically every area of the economy and is an integral aspect of a country's economy. Artificial Intelligence is no exception. Businesses have embraced practically all developing technology in their software development. AI's effect on software development alters how businesses operate and makes software smarter. In this paper, we have talked about the change AI would bring to software development and the role of could. We have also discussed traditional ways used. Towards the end of this paper, we have added open issues and challenges in detail.

Chapter 4 talks about education and its relationship with Industry 4.0. The technological advances that the world has experienced throughout the years have caused important changes in the structures of society. Education is one of the fields in which the industrial revolution, driven by technological development, caused a significant impact since technology allowed students to freely access more information and therefore more knowledge. Education 2.0 and 3.0 brought not only access to information but also greater connectivity and interaction between people, particularly between teachers and their students. The dynamism of technology also brought opportunities for innovation with Education 4.0, changing the existing paradigm from face-to-face to online classes, the emergence of digital, your relationship with those around you, and your abilities to meet basic demands and roles. in society. These changes meant the application of new educational methodologies, aimed at significant learning that prepares the individual with the necessary skills to face the modern world of work.

Chapter 5 reviews and discusses three pervasive topics in current manufacturing processes around the world: 1) Internet of Things, considered a key factor in the Industry 4.0 revolution, and how its applications can be used to enhance supply chain and manufacturing operations; 2) Serialization and Tracking of Manufactured Products, which discusses the value and usefulness of product traceability for warranty purposes and to reduce counterfeiting; 3) ERP systems in manufacturing processes, which, when used appropriately, provide real-time data and analysis capabilities for instant and long-term decision-making, which in turn can provide a competitive advantage for companies. The goal is that this chapter may serve informational and referential purposes, and as a practical guide to either launch or implement real-world careers or processes in the areas discussed.

Chapter 6 deals with blockchain. Digital technologies such as IoT, Cloud, and mobile computing have enabled product innovation, consumer interactions, and efficiency. Blockchain technology has altered how organizations work, providing a paradigm shift in resilience, security, and efficiency. Blockchain enables agile value chains, product innovations, and technological fusion. Every company considers research and product development expensive. Investments are risky, especially if the new product version differs considerably from their current portfolio. Firms must share high investment risks through joint ventures or cooperative product development. However, for collaborative product development, trust and transparency must be vital. Decentralizing design data interchange, funding, and marketing with correct crediting of contributions is one of the methods for multiple-organization collaboration. Blockchain technologies can be employed for such collaborations. This study explores how blockchain can decentralize product development worldwide to meet security, scalability, efficiency, and incentivization criteria.

Chapter 7 deals with Industrial Cybersecurity. Cyber threats against industrial manufacturing are escalating as more and more connected systems create larger attack surfaces. Being Cyber-attacks one of the top concerns to interrupt operational continuity, manufacturers must establish a complete program to protect, identify, detect, respond, and recover. Highly accurate detection of cyber threats, risks, and anomalies improves operational resilience, no matter the age of the facility or the scale of the production lines. In terms of industrial cybersecurity, there are specific challenges to take care of: firstly, Legacy Manufacturing Devices Were Designed for Reliability, Not Security; secondly, Network Segmentation and Zero Trust Have Been Slow to Come to Manufacturing; and exponential growth of IoT Sensors.

Chapter 8 deals with bibliometrics. Cybersecurity is one of the main concerns of most organizations, derived from the violation of their cybersecurity. The objective of this article is the development of a bibliometric study on cybersecurity from the organizational level. For the method, the Web of Science platform was chosen, since it indexes and organizes scientific production and has the high impact factor of the

Journal Citation Report (JCR). The results show evidence of the evolution in which cybersecurity has been addressed within the organizational field, so the originality is related to the business importance in this globalized world. The various theories and approaches used in empirical studies provide valuable perspectives to understand the behavior of the actors involved, the motivations behind criminal actions, and cyber risk prevention and mitigation strategies, contributing to the good performance of organizations. These findings show the interest of the scientific community, observing a growing trend around publications within the subject of study.

Chapter 9 explains the technology applied in solder equipment. The industry has raised its operations by manufacturing different products for a certain purpose, electronic assemblies have been the focus of the development of panel circuit boards (PCB) through different processes until they complete an assigned product that accomplishes the specs required by the customer and comforts their life. A main and useful technology implemented for mass PCB soldering in through-hole areas; is the Wave Solder equipment, their usage provides to industry the key to accomplish with daily demands in a lower time compared with manual process costs.

Chapter 10 deals with CPS. By employing a constant flow of information from the production processes, manufacturing Cyber-Physical Systems (CPS) is a step ahead of industrial automation from a traditional manufacturing system to a well-networked and adaptable manufacturing system. By learning from the data obtained from real-time and non-real-time production processes that can operate independently, this system can optimize itself and adapt to the changes. With the advent of IoT and CPS, a connection between machine to machine and from machine to human has been established in industrial communication systems, resulting in more efficient and flexible systems that have decreased production time and improved forecasting capacities. CPS focuses on how physical systems and computing networks can communicate. It is applied to IIoT, cyber security, and smart manufacturing. This paper aims to provide an overview of CPS deployment in manufacturing. It discusses CPS development, CPS adoption, and CPS requirements for Industry 5.0. Finally, the main CPS production problems are also discussed.

Chapter 11 is an exploratory analysis is presented about the salaries granted by the maquiladora industry in Tamaulipas and its effect on the poverty indicators, with the claim that the models of linear regression always determine the consistency in both sign and statistical validity, as well as the magnitude of the impact. Five relationships were raised through the Ordinary Least Squares, finding that, the coefficient does not meet the expected sign, that is, it should in terms of the intuitive logic have a negative sign, given that the interpretation is that the increase in salaries causes an increase in the levels of Poor People, People with moderate Poverty and Vulnerable by Income. The only relationship that satisfies both the sign and the individual significance is that People with Extreme Poverty indicate that the salary allows them to get out of that level. It is suggested that additional observations must be collected throughout time, given the lack of these available, to validate the proposal.

Chapter 12 deals with salary in the manufacturing industry. In today's global economy, emerging economies play a very important role in the production chain. Countries such as China and India have become competitive participants in the global market.

The labor force in these emerging markets has developed new skills, leading developed economies such as the United States to move important areas of their production chain into these countries.

This phenomenon had reached Mexico as well and multiple cases of salary and wage inequality could be found.

Thus, this research tries to establish a frame that could help better understand the effect that salary and wage inequality play in the leadership-productivity relation among middle management supervisors within the manufacturing industry at the Mexico-US border. A questionnaire was developed and applied to middle management supervisors on both sides of the border (Mexico and the US). The main contribution of this study is to help better understand the impact of salary and wage inequality among middle management supervisors.

Chapter 13 aims to define the importance of support committees in the manufacturing industry. Emerging technologies, understood as new technologies or the continuous development of existing technologies, have led to significant advances ranging from the simplification of everyday tasks to the automation and optimization of industrial processes; and therefore, to increase the productivity and quality of life of people. Who does not like to have more time to devote to other activities or to rest, or to have some device that allows one to simplify life with a click? Or in the case of companies, reduce production times, costs, and process failures. While we may think that responsibility for promoting the progress of emerging technologies; as well as the definition of public policies, regulations, and regulation of this matter lies in the figure of the government, the reality is that the work and interference of other actors such as the private sector is crucial to its progress.

Finally, Chapter 14 talks about humans and their relation with the business environment in the new reality. Technology tools have played a key role in the transformation of learning and remote work, as well as, in employee retention. Through access to information, improved communication, and optimized productivity, these tools have enabled organizations to meet the challenges of the digital age and flourish in an ever-changing environment. However, it is important to remember that technology is only a tool and that its true value lies in its effective integration into an overall strategy that promotes people's engagement and well-being. By embracing the potential of technology in a humanized way, organizations can build strong teams and successful projects in today's digital world.

The book constitutes the first effort between The National Council INDEX National, along with its 18 associations, representing the productive interests of the Maquiladora and Export Manufacturing Industry. INDEX is the most socially engaged industry, due to opportunities for more Mexicans to enter the world of manufacturing and exporting, all while maintaining social balance and spurring national and regional development. INDEX National is advocating for an industrial policy that extends to more sectors and encompasses all those with an exportable focus. In this context, Twin Plant (IMMEX) companies are emerging as pivotal players due to the escalating importance of global value chains.

And in the other hand, the Autonomous University of Tamaulipas is a Mexican public university based in Ciudad Victoria, Tamaulipas. Throughout the larger cities of Reynosa, Matamoros, Nuevo Laredo, and Tampico and smaller cities of Ciudad Mante and Valle Hermoso, offering undergraduate studies. Each of the various university faculties offers graduate studies leading to a master's or Doctorate degree. The university's Center of Excellence sponsors specialized professional certificates and studies for the state of Tamaulipas.

This joint effort brings the first edition of *Emerging Technologies and Digital Transformation in the Manufacturing Industry.*

Fernando Ortiz-Rodriguez
Tamaulipas Autonomous University, Mexico

Sanju Tiwari
Tamaulipas Autonomous University, Mexico

Luis Hernandez
INDEX, Mexico

Agustin Tiburcio
INDEX, Mexico

Chapter 1
Reshaping the Role of Finance: What Matters Most After the COVID-19 Pandemic in 2020

Luis Manuel Hernandez-Gonzalez
INDEX, Mexico

ABSTRACT

The world of manufacturing has shaped several countries and regions, and created specific sites of excellence in several places. However, the speed at which investment is being made requires all support areas to be agile and effective on the deliverables. Finance plays a significant role from day-to-day activities, to strategy, investment, and in the relocation of manufacturing footprint. The speed of cost, rate of revenue, and inventory have become a major challenge for companies and for finance executives.

INTRODUCTION

Covid-19 pandemic in 2020 (Nasereldin et al., 2020) gave humanity a new perspective of life, time, business and how to prioritize what matters most for each of us. As a result of these radical changes, business came into a conclusion where the reorganization of supply chains was acceptable for a world before the pandemic; we all suffered the face masks shortages, spike in commodity prices and identified that cost was a matter with less leverage that time and availability of goods.

We saw domestic challenges, international challenges, and identified that countries where not ready for a new reshaping of supply chains, inventory management, time to market (Medina-Quintero et al., 2023), market share shifting from some areas to other and a threat to economy as we are now it and, in some cases, national security.

DOI: 10.4018/978-1-6684-8088-5.ch001

The manufacturing industry is one of the most relevant economic contributors for many countries, such sector has been growing in skills, automation and as well has been the trigger for some of the most relevant changes in trade regulation and how economies plan their budgets around the globe, such as:

- International trade regulations
- Standardization of global regulations such as ISO, FDA regulations, Aerospace, among others
- Ports development and logistics standards
- Workforce development
- Assets investment and asset management for companies
- Labor arbitration
- A key index of countries industrialization, education, and GDP
- Among others

As an example, one of the most relevant trade agreements is the USMCA in which the United States, Mexico and Canada define the rules of the ground to manage commerce within them. The USMCA daily trade is estimated to be $3 million dollars a minute with 2022 trade data. Such a large figure includes material costs, indirect costs such as labor and support and utilities and other fixed costs. Companies in the USMCA region, such as any other trade agreement, have created companies that, despite the sector, need to be resilient, sustainable, innovative, and scalable; so, manufacturing technologies have shaped economies and will reshape them in the next 5 years. Furthermore, we need to assess in those companies Environmental, social and governance (ESG) responsibilities; this creates a manufacturing system that will need to be managed differently.

STATE OF THE ART

Finance and accounting are areas that have been evolving at a slower speed than manufacturing, such lag is due to regulations, having accurate financials take deep analysis and numerous of iterations to reflect accurately what the company did and will be doing; and if a company is publicly traded there is a need to comply with several SEC, IRS, and other agencies standards. Despite this, we need to understand that the finance role is a key element on any manufacturing site, if the finance executive understands and acknowledges that there are gaps in the level of information that is managed in the shop floor is different to what is managed in their accounting records. I am not talking about general ledger only, I am talking about cost accounting, inventory management, payroll costs, manufacturing overheads, allocation of cost, sales and cost of goods sold, in other words linking the shop floor to the financials has been difficult. There has been great progress in the financial function applying lean techniques, and one of my favorite books on this topic is "Who is counting?" by Jerrold M. Solomon (2003), which provides us with a high-level view of how a finance executive gets into the world of manufacturing, lean manufacturing, six sigma and other interesting projects that are day to day activities in quality, materials, engineering and other areas.

Manufacturing costs include normally Material, direct labor and Overhead costs, while this chapter is not intended to reflect what is included in each of them, we can easily relate to material as the raw material used for production, the headcount used to manually or automatically process is considered direct labor and costs such as lease of space, depreciation, salaries, utilities are considered overhead. In the manufacturing sector finance works based on and accumulation of data normally in software or spreadsheets; then that data is coded in specific accounts and with that information we decide what to do.

Based on the article called "Manufacturing process innovation for industrials" (Fletcher, 2021) in the year 1947, manufacturing accounted for 25 % of United States GDP and by 2019 it was down to 11% of country GDP.

A reason of this shift has been triggered by the exit of intensive labor operations out of the United States in search of lower cost, as a result there was a reshape of the US manufacturing footprint, having large scale operations in countries such as Mexico, China, Vietnam among other. With the pace of automation and recent technology developments, the new norm has created smaller operations with more capabilities, flexibility, error proofing equipment and a more specialized worker.

The automation process in support areas such as materials, quality, human resources, finance, logistics and several more has not been as fast as in the shop floor for manufacturing, we have seen development and as well seen integrations in ERP's and several other forms of software. Specifically in finance we have seen great automation on repetitive processes such as accounts payable, accounts receivable, taxes, payroll, general ledger to mention a few. We could argue that this is most of the finance function, which might be true; however, a few of the information automated on these processes can support operational decisions. Some of the most relevant elements of information's for manufacturing would be orders management, sales, cost of goods sold, EBITDA and profit; having this information handy can provide management with relevant insights on day-to-day activities and have a certain accuracy on forecasting the month, quarter and with this information define either the strategy or the tactical efforts to be released.

RESHAPING OF THE WORLD SUPPLY CHAINS

After COVID-19 pandemic we identified that supply chains (Baldwin, 2012) where something we lived and thought we understood, however many of us were looking for facemasks at a certain point of time, and availability was several times referred as "they are in a container in China"; so, the end consumer understood that inventory management was something to consider in our day to day lives. Another of the shortages we lived through was chips and as a result we could not purchase a television, or other appliances at the time we wanted them. To understand the challenges and later analyze the manufacturing actions and solutions we could use; I will refer to the article from World Economic Forum article "5 Challenges facing Global Supply Chains" (Meixell & Gargeya, 2005). as follow:

1. The rising cost of living
2. Labor unrest
3. Energy Shortages
4. Geopolitical uncertainty
5. Extreme weather

Related to manufacturing and in order to understand the rising cost of living we should think of this as an economic impact in our operation, this is that due to shortages in materials, due to demand increases we started seeing an increase or inflation in prices, somehow triggered by market reactions and consumer change of habits, this process generated uncertainty in producers and sellers, that made most of them increase prices to maintain continuous flow of goods, or to maintain profit or to keep inventory on hand as a result of not being able to predict demand. This inflationary increase triggered increases in labor costs so the consumer could keep their spending power and get the goods in the market. "Energy shortages has been seen in several border towns, as an example of gas shortage due to the Ukraine war, gas shortages in the area of Texas, USA and Chihuahua, Mexico border, in which due to low temperature in 2021 the government in Texas decided to stop the supply of gas to the border town of Juarez, which created a cost increase of the gas in Mexico and US consequently; also this created supply shortages."

SHOP FLOOR MANAGEMENT

Working on the shop floor of a manufacturing site is one of the most dynamics roles we can see, there are either emergency, situations, challenges and is where if there is a tight correlation between all support areas it will deliver the finished goods as expected, if not, then we will see a plant under perform and as a result their optimal financial performance will be missed. Shop floor is normally working with up-to-date metrics, most of them related to quality, delivery, and output, with those metrics' decisions are made and supported by most of the teams; with this the operation can compare if the day was good or bad; either we made it or not.

For some of us, finance is seen as an area who accumulates data, analyzes, provides insights, and shares what was the result of decisions made. We could argue as well that this function is called a lagging indicator, meaning that we need to have data to measure the result or after the fact measurement. With the Covid-19 pandemic at its highest peak in the year 2020, several organizations required many anticipated financial models and with most of us isolated from our offices, factories, and warehouses, calculations were made based on historic financial data and not performance metrics tied to financial data.

SHOP FLOOR COST FORECASTING

Before COVID-19 financial forecasting was a regular exercise for several organizations, but as of today financial forecasting has become a short term strategy tool that requires from the finance function better knowledge of what is being made, how is being made and the resources needed for this, below we can identify the top questions we must address in the main financial statements, ***Profit and Loss, Balance Sheet and Cash Flow***, we can use this as a framework of what can be used to create better assumptions and create the check and balance actions to continue on improving forecasting accuracy. The below categories are the main ones used through several industries and do not provide a fully detailed list of actions; it provides a roadmap for users.

Table 1. Profit and loss statement: Sales and revenue

Category: Sales growth/decline Sales
Key Market considerations 1. Are the sales based on current market growth/decline? 2. Are we taking sales from our competitors, or are our competitors taking sales from us? 3. Is there a trend in our market that we are not aware of for the growth/decline in sales? 4. Are we selling more/less because our competitor changed the price? 5. If we have distributors, is the rise/decline in sales a result of their good/poor inventory management? 6. Is the customer buying more inventory to exit our business? 7. Which sales are being triggered by contract requirements and which ones are being triggered by the market? 8. Are our sales impacted by any financial currency changes, either increase or decrease? 9. Are our foreign sales driven by either geoeconomic or geopolitical factors?
Key Capacity considerations 1. What is the cycle time to build the product? 2. Do we have the inventory to build the product? 3. For how long we will be expecting the same number of orders to come in? 4. Does our current capacity support the sales requirements? 5. Do we need to hire more personnel, and by when we should stop hiring personnel? 6. In terms of capacity, how much is productive capacity, how much is non-productive capacity and how much is available capacity. 7. Do we need to invest more in Capex or Opex to support the current sales increase? or do we need to shut down shifts/machinery to support sales decrease? 8. With this sales increase/decline, do we need to change the maintenance schedules for assets? 9. How much of my depreciated assets are being used to build these sales?

Table 2. Profit and loss statement: Material costs

Category: Material costs
Key Market considerations 1. Where is my main source of material/inventory located? 2. What is the currency telling us, are we having gains or losses by currency rates? 3. Are the market prices increase/decrease absorbed by our customer, our supplier or by our company? 4. How much material on hand does our supplier have in their locations? 5. Are the recent changes in tariffs, duties, or taxes that we need to be aware of and notify our suppliers/customers? 6. Are there any market shortages of materials that could impact on our supply base? 7. Is our supply rating program providing us with any intelligence that could lead us to inventory shortages? 8. What is the lead time for us to get material from the supplier to our warehouse? 9. Are there any new market players that could hurt our supplier chain, by either price, location, quality, or compliance matters?
Key Capacity considerations 1. What is the cycle time required to build the material we need to fulfill our sales? 2. How much material on site and on hand do we have to fulfill the sales? 3. What is my current cycle count/inventory accuracy in our warehouse? 4. Are we purchasing material based on bill or material usages, and how accurate is my bill of materials? 5. Do we process shop floor material discrepancies in time or are there any pending adjustments to be made? 6. For the sales increase/decrease do we need to release some storage capacity or require more storage capacity and for how long? 7. Of my total inventory, which components are unique, and do we have the required amount to support production rates? 8. Of my total inventory, which components can be used as alternate material with no major disruption on the customer side, which would allow me to better control/reduce our inventories? 9. Of my total inventory, how much is obsolete and excess, do we need to have a disposal event that will lead us to purchase additional material?

Table 3. Profit and loss statement: Direct labor costs

Category: Direct Labor
Key Market considerations 1. How is the market reacting to direct labor cost increases? based on performance, merit increases or inflationary considerations? 2. Are we getting labor arbitration benefits in some of our sites or countries? 3. Is there any tax rate adjustment to consider? 4. Is our union aware and aligned with the market? 5. Are there any benefits changes we need to consider? 6. Is the tenure of the employees aligned with the market so we can compare salaries? 7. Do we have more tenure employees that do not allow us to quote lower cost? 8. Does the direct labor market turnover is based on new
Key Capacity considerations 1. What is the direct labor market turnover right now and what is expected in the next 3 months? 2. Do we have a complete roster to build the required demand? 3. Which day of the week is the most productive day? 4. Is my time to build the product aligned with the direct labor associates we have on the shop floor? 5. Would there be any risk of requesting overtime for employees? 6. What is the best shift we have and how does that potential our capacity? 7. Which is the best manufacturing cell productivity rate and how can we replicate it? 8. How do we quantify the idle time employees have from current scheduled maintenance and non-scheduled maintenance? 9. Will there be any benefits from having a lower/highest cost of employees for building the current orders request?

Table 4. Profit and loss statement: Overhead cost

Category: Variable and Fixed Overhead
Key Market considerations 1. Are we getting any currency gains/losses in our cost we need to be aware? 2. Are there any contract changes regarding costs that we need to consider? 3. What does the expected inflation increase cost in utilities? 4. Are there any taxes such as property tax, wasted disposal, that we would need to consider? 5. Are there any inflationary changes we would need to consider for Variable overhead suppliers, such as MRO costs, maintenance, repairs, etc.? 6. Will there be any employee benefits change that we would need to consider? 7. Are there any costs we would need to adjust up/down due to market shortages or demand changes? 8. Regarding leases, is there a change in methodology/accounting that could make us miss our target? 9. Are our competitors making changes because of cost increases, such as pass through customers, or suppliers?
Key Capacity considerations 1. How many months of MRO products do we have on hand? 2. If the demand goes down/up, are we adjusting down the open purchase orders or blanket purchase orders? 3. Are we changing our minimum and maximum levels for supplies, and spare parts purchases as volume changes? 4. Are there any open purchase orders we need to close or cancel? 5. Are there any pending goods to be received that might impact our financials? 6. Regarding fixed assets, how much of the assets used for production are assets totally depreciated and how much are assets still depreciating? 7. Regarding amortization costs, how much of what is used for production is totally amortized and how much is still being amortized? 8. Regarding rent of space, how much is being utilized by true production floor, how much for storing material and how much for offices? 9. Regarding salaried and indirect employees, how many are used for production and how many are used in support functions?

DRILLING DOWN ON COSTS

Financial leaders need to drill down costs and avoid the systems and reporting traps, while a spreadsheet or software can provide insight, hunting down true root causes, cost and understanding the behavior of them in different scenarios. We can use the "cost driver" analogy used back when ABC costing was new, now we need to experience beyond the fragmented point of view from cost centers. We need to move to an E2E (end-to-end) cost understanding and for that we need to understand how the company solves problems for customers and what customers are willing to pay for those problems solved. For these reasons I identified four main factor to drill down costs before we get in trouble:

1. Drilling Down cost on Suppliers: This includes understanding our suppliers' behaviors, what they do with their waste, how are they productivity improvements translated in more business, how do they manage their inventory can tell us as well how we will manage ours. If we can understand how they use their assets a lot of our problems can be understood.
2. Drilling Down costs on Design: While designing we learn a lot from the manger that how to drive into a market or a manager who is clever enough to create chaos so he can fix stuff. Design costs have a direct relationship from how a shop floor or manufacturing site is working, despite the material composition of cost, the efficiencies and productivities in the manufacturing site will need to be tested in the design of new products.
3. Drilling down cost in Supply Chain: This one by itself provides a lot of information and requires the same level of analysis, however, the cost incurred in the supply chain can be analyzed on how much time in the supply chain is the product, as an example: if we carry goods from an outside location we can calculate the days in transit and how much we paid for the time each of the providers took our product, we can calculate that maybe that the carrier cost is lower when comparing time in transit from the source to the required place versus local carriers that are just moving material for storage.
4. Drilling Down on Process Waste: Waste is related to what we understand as lean manufacturing, often I would follow the approach of eliminating waste rather than cutting costs; this is an exercise we must engage normally in support areas, rather than operational areas, this is because support areas need to provide more often true cost savings rather that cost avoidance efforts, or what is know as soft savings, we must remember than when we start cutting costs if we do not eliminate waste first we could hurt our customers by affecting product quality or other functionality of finished product.

We all can improve our financial acumen and knowledge, however a lot will be coming from what we can learn from the people on the ground, the people that is doing the product, the people on warehouses, shop floor, quality areas, and by experience I can say that learning what to ask to them will be the best experience provider for any manager, director, financial leader or general manager.

THE NEARSHORING EFFECT AFTER COVID-19

We learned that supply chains are what can make a region strong or weak, with the mine-to-end strategy that China has developed, several of the raw materials will be managed by that country and their allies, along with the suppliers controlling it and providing first to the Asian market will be their top priority. With this in sight, the United States is targeting on getting a closer to home supply chain and we have heard the buzz word during 2022 and 2023 which is: Nearshoring. The sole purpose of nearshoring is to have a supplier close to the market and preferably within a trade agreement with lower duties and taxes. The nearshoring effect for United States will be twofold:

(1) By moving either united states companies in Asia, closer to the USMCA region or
(2) Asian companies moving small portions of their operations to Mexico to get the benefit of lower duties by using Mexico's footprint to qualify as USMCA origin products.

Within this framework manufacturing companies could potentially increase demand and will need to understand effectively how to the get the most of their capacity, as well, capacity constraints within the sites will need to be evaluated since we would need to understand how productivity plays a big role in getting some of the nearshoring opportunities. According to Morgan Stanley research document "Mexico is poised to ride the nearshoring wave" [6] "If U.S. manufacturing is to be less dependent on China, we think the path will be via Mexico," says Morgan Stanley Research equity analyst Nikolaj Lippmann. "Nearshoring is expected to be a long and sustained race that could help build new ecosystems in Mexico's existing manufacturing hubs." This will trigger a growth of $154 billion dollars on the following markets.

- $94 billion in gains in well-established sectors such as electronics and automotive

- $38 billion from sectors that have benefited from free trade via the United States-Mexico- Canada Agreement, or USMCA
- $22 billion from a second wave of nearshoring growth driven by increased manufacture of IT hardware and new opportunities related to the electric vehicle supply chain and other clean technologies.

CONCLUSION

Nearshoring will be the game we will be playing in the following 5 years, this alone needs to be a focus for manufacturers, suppliers, carriers, schools and regions. Becoming successful in securing the nearshoring moves will be a catalyzer for new economic and trade regions, we will be able to see different new sectors and investments in the year 2030 and as a result there will be new regional economies, changes to infrastructure and a change in the vocation of the population; we must remember that capital and talent move where they have better options to grow and access what they want.

REFERENCES

Baldwin, R. E. (2012). *Global supply chains: why they emerged, why they matter, and where they are going.*

Fletcher, R. (2021). *Manufacturing Process Innovation for industrials Manufacturing*. McKinsey.

Medina-Quintero, J. M., Sahagun, M. A., Alfaro, J., & Ortiz-Rodriguez, F. (2023). *Global Perspectives on the Strategic Role of Marketing Information Systems*. doi:10.4018/978-1-6684-6591-2

Meixell, M. J., & Gargeya, V. B. (2005). Global supply chain design: A literature review and critique. *Transportation Research Part E, Logistics and Transportation Review*, *41*(6), 531–550. doi:10.1016/j.tre.2005.06.003

Nasereldin, Y. A., Brenya, R., Bassey, A. P., Ibrahim, I. E., Alnadari, F., Nasiru, M. M., & Ji, Y. (2020). Is the global food supply chain during the COVID-19 pandemic resilient? A review paper. *Open Journal of Business and Management*, *9*(1), 184–195. doi:10.4236/ojbm.2021.91010

Solomon, J. (2003) *Who's Counting? A Lean Accounting Business Novel*. WCM Associates.

Chapter 2
The Foreign Trade Software Control:
Effects of the Deficiency of Its Sustaining and Challenges in the Replacing

Pedro Ramirez
IIES, Mexico

ABSTRACT

There is a much biger world behind a company's import or export transactions than meets the eye, and understanding the inferences that suffer by not having proper administration through customs software is important. Being a company of the IMMEX program in Mexico is critical, and when there is the need to change software, a company will face a series of challenges with which they will have to demonstrate the administrative skills of the project leader and the rest of the personnel involved.

INTRODUCTION

When you have worked in the sector of the "maquiladora" industry and talk about the software that the company where you work has implemented, what surely comes to mind, as the first option is to think about the MRP (Materials Resource Planning) or the ERP (Enterprise Resource Planning). The ERP that controls manufacturing and surely other areas as well; Finance, Human Resources, etc. However, if we ask you about the foreign trade software, it is very likely that you will not answer which one is in place, unless you work in the department that manages it or that interacts

DOI: 10.4018/978-1-6684-8088-5.ch002

indirectly with the system. Well, a Customs or Foreign Trade administration system is specifically for this management. Not every ERP includes the administration of customs transactions and companies know this, so they usually go for a specialized system to perform these functions, more thoroughly, if the ERP that you have throughout the plant can be linked to the foreign trade software can always be considered as an extra benefit.

In the North side of Mexico, the fact that a company has import and export activities implies a responsibility in which it must immediately have a software that manages those activities. Moreover, that is that in the northern border of the country, manufacturing companies have increased enormously and with it the use of this software, in this way Mendoza J. mentions a relationship with the implementation of NAFTA:

For the Same Reason, it is highlighted that since the establishment of NAFTA there has been a notable increase in employment in the manufacturing sector of Mexico. In the first ten years of the treaty, around one million jobs were accumulated; however, this statistic did not have a significant impact on the dynamics of total employment in the Mexican economy (Blecker, 2014). As a complement to the growth of manufacturing exports, manufacturing imports also had a very dynamic growth, which was concentrated in the import of supplies for manufacturing export production. (Mendoza, 2022, p. 4)

The relationship that exists based on the growth of export manufacturing industries and the country's economy increasingly requires more professionals specialized in foreign trade and therefore software that helps and supports all the functions required by movements through American customs for exports and Mexican customs for imports. The export manufacturing industry has become a fundamental sector of the economy of the northern border region, both because of its influence in total exports and because of its contribution to job creation (Comercio exterior y empleo en la frontera norte de México, 2022, p. 4). Hence, it turns out to be an unavoidable problem since the more imports and exports; more precision is required in the software used by the companies. If it is not enough and the problem persist, it can be solved by updating, but this leads to another list of problems because the transition from old system to new system (Medina-Quintero et al, 2022) . The transition can be as complicated as the lack of control that the company has in managing its import and export transactions. The program that is used may or not be linked to your ERP automatically, but the integration between these must be mandatory because in terms of scope the system that is utilized must even support audits by the Mexican government. That being said should be clarified what; a software does not replace the professional, conversely, it requires someone with the knowledge and bases of local and international regulations.

This work is narrated based on the author's experience, citation from research journals and some participation from professionals in the area of international trade. Given the need for a software change and given certain factors that influenced when a manufacturing company in the aerospace industry is involved requiring a change of the customs manage system and the implementation of a new one in a setback. As well as explaining in a simple way, the relevance and influence that software has for foreign trade activities, and answering regarding: What advantages can it bring having a software for customs and tax management? What is involved in such a software change? Under what unique circumstances can you face difficulties when switching trading software whatever the commercial business is?

FRAMEWORK: STATE OF AFFAIRS

For just over than two decades, a company dedicated to the manufacture of high-performance semiconductor components for the Aerospace branch, has as a manufacturing plant in one of the most important border cities in the north of Tamaulipas. This plant has in its product portfolio, analog and mixed-signal semiconductors with exclusive high performance, offering catalog products and its list of qualified products, QPL (Qualify Product List) and also products exempt from this list (Non QPL), rectifiers and transient voltage suppression products for a variety of customers

The commerce of the company and the nature of the product make the infrastructure of its facilities particularly peculiar in terms of size, moreover, the profitability of the product has kept it active for years, giving way to jobs with very minimum turnover., employees with too much seniority who enjoy working in a safe, stable environment and with satisfactory remuneration. The dimensions of the products make that packaging and shipping are of smaller volumes compared to almost any other company in the region. Other companies that use pallets, forklifts, storage racks with four or five floors, and warehouses with large amount of personnel, but, nevertheless, by producing products of dimensions close to those of a shirt button, all these tools are unnecessary in the flow process of the plant and for the storage of the raw material or finished good parts. Since this is an international company, it has to import the raw material from the United States with a frequency of one to two times a week and when the raw is manufactured, has to be returned already transformed as a finished product, exporting twice a week.

From the point of view of customs legislation, the amount of imports and/or exports that a company carries out does not determine whether to declare or not declare to the Mexican government agency. But rather the contrary, since an importer is defined as: as the individual or company that wishes to import merchandise into national territory according to the and this action is regulated and controlled in tax matters to a merchandise of foreign origin to allocate it to an economic function of use, production or consumption (Tributaria, https://www.sat.gob.mx, 2018) Likewise, any export is explicitly mentioned to be operation through which national merchandise is sent abroad in a legal manner regulated by tax and customs laws to be used for an economic function of use, production or consumption (Tributaria, https://www.sat.gob.mx, 2017). No company that carries out customs operations is excluded of the required compliance.

For customs management activities, companies generally have exclusive software that is responsible for all integration and compliance with the country's public agencies, such as SAT (Tax Administration Service), SHCP (Ministry of Finance and Public Credit) or SEM (Mexican foreign service). However, for a software to achieve its maximum potential it specifically requires its updates, the inputs of information so that it can generate deliverables, the integrity of this information that is loaded, the maintenance of source code, and possibly the main part is the update. This update of the software has to be aligned to the requirements of a changing customs system and the bureaucratic regulations that the government system frequently launches. Consequently, the systemic part is complete if an organized operation is carried out, but it cannot be taken for granted that the software itself will do the work or follow up on everything that foreign trade needs. Therefore, here comes the human resource in the Logistics department or also known as Imports / Exports thus is abreast in importance than a software. Since the level of competences that the person in charge must meet, it is necessary to have an ad hoc profile. The leader of this needs the expertise to be at the forefront of new legislation and handle the bureaucratic online and/or face-to-face systems as a normal process. If it is not possible to sustain all this, the software and processes that are in place could eventually stop being functional and customs operations collapse due to the lack of continuity in compliance and updates.

In this company, having a reduced number of movements through customs, the operational part, which is understood to be the one who directly executes the tasks and interacts with the surrounding departments such as: receipts area, quality inspection, shipping department, etc. It is still important but not justified enough in order to have many number of associates. The software used for customs management has been the same for more than ten years for which a monthly license is paid to update the code itself but not what is required by government changes (Ortiz-Rodriguez et al, 2023)

We Have Always Done It This Way

The Import/Export department has only one operational person who reports directly to the Purchasing and Materials Manager. The daily activities that the plant requires, such as: shipments of raw materials to sub-processes with external suppliers, shipments of finished products, commercial invoices submitting, declarations, etc., have been carried out in a regular and monotonous manner for many years. These have been adjusting parameters in the software only when required doing this through a ticket requesting support from the technical team of the company that provides the Software service. The monthly, quarterly and annual requirements are executed with the support and help of companies that function as a customs agency and the intermediary commonly called "Broker", these perform certain limited functions, extending the support beyond their duty and sometimes out of their scope, from the customer-supplier point of view. Besides, each delivery period or each customs report that has to be provided using only estimation based on historical data, past experiences, and in some measure with premeditated, arbitrary, and risky actions are always used in order to resolve the situation and comply with what is requested. Given a lack of basic knowledge due to not having the bases of the customs administration, the profile of the current person who performs the operation is not enough to carry out such tasks that require knowledge of compliance, periodic obligations, and the responsibilities that each government agency has. Thus, the participation of the manager department ends up being key, since by the simple expertise of investigating and asking questions, each situation has been resolved overcoming obstacles unnecessarily. In addition, it is running the risk of fines and greater consequences, but by always doing things like this, you cannot see the magnitude of the severity and spend the years facing the same questions every time they have to comply with the customs law. Then; is it possible that a company can subsist without updating or organizing in its customs transactions? The real world is that, running jeopardize is something that nobody wants, government audits are impartial and strict, so you might as well think about who wants to be fined and pay for something that could have been avoided. Talking about the understanding of the risk that is run by not understanding what to do things because that is how they have always been done; we could mention an example: during the end of the year where the Broker notified that, we were about to lose the IVA / IEPS certification. Since the time of the Payment and delivery of paperwork was due and the follow-up to this was not properly carried out, the managers of Finance, Operations and Materials met with the site Director to see the alternatives. With time over and the December holidays just around the corner, the payment was not made on time, and the SAT office began its vacation period, leaving on hold all the procedures that did not fall within the established time. The delivery of the information was more complicated since the system that was managing customs issues was outdated, with imprecise information to show the balances of raw materials, equipment, etc. By then, an intense search had already been made for the latest information that the previous Materials Manager on charge had done and as it had been presented, the periodicity, the documentation, the instances, the factors, everything per se, was completely unknown. There were proposals and ideas, but with vague backgrounds, since there was no expert person and obviously, the software was imprecise, the lack of system maintenance exposed us more and once again highlighted the "it's always been done that way."

At this point, the Unique Notice of Renewal formats in the Registry of the Certification Scheme of Companies was the option to obtain a registration again. It is because there was no confirmation from the SAT since they did not return from vacation until the beginning of the year and by then it was too late. It was decided to hire an external consultant from Mexico City (CDMX), was whom followed up, and to whom all the documentation was, gave it in order to maintain the registration and to grant recertification to the program. Through the consultant and the broker who worked focused on the issue of the company, was possible to maintain the IMMEX program at the end and could continue working with the IVA and IEPS tax credit.

TAKING RISKS: THE RISKS ON THE IMPORT AND EXPORT ACTIVITIES COULD BRING SERIOUS PENALTIES

Working an operation of whatever magnitude or infrastructure is, under the conditions of "we have always done it this way" has a high risk of fines such as those published in the Diario Oficial de la Federación DOF: 12/24/2021 (Sánchez, 2021). Which mentions, among others, articles 184 and 185 of the Customs Law, and refers to those who commit an infraction related to the obligation to keep records and inventory control systems could be fined in amounts greater than twenty thousand pesos and mentioning that these are applicable from January 1, 2022.

In accordance with the provisions of the second paragraph of this rule, in Annex 13. "Updated fines and amounts established by the Customs Law and its Regulations". The updated amounts are disclosed, effective as of January 1, 2022, from the amounts established in articles 167-D, section I; 167-G, section IV; 184-B, sections I and II; 185, sections I and XIV; 187, section IV, of the Law. The aforementioned update was carried out in accordance with the following procedure: a) The latest update of the amounts established in articles 184-B, sections I and II; 185, sections I and XIV and 187, section IV of the Law, were disclosed in "Annex 2 of the General Rules of Foreign Trade for 2018", published in the DOF on December 24, 2018, which entered into force on January 1, 2019. For the calculation of the update, the accumulated percentage increase of the INPC in the period from November 2018 to June 2021 was considered, which was 10.47%, exceeding the 10% mentioned in article 17. -A, sixth paragraph of the CFF. Said percentage is the result of dividing

113,018 points corresponding to the INPC for the month of June 2021, published in the DOF on July 9, 2021, by 102,303 points corresponding to the INPC for the month of November 2018, published in the DOF on December 2018, minus the unit and multiplied by 100] (https://www.dof.gob.mx/, 2021)

In our country, as in any other, the companies with customs activities and transactions; importing and exporting, they have a key place in the development of their city, state and country and that implicitly is the growth and success of the companies. However, would be severely risky not having foreign trade software and the impacts are expensive that could negatively affect organizations as a result. With this context, the company always ran the risk in commercial operations, with a high probability of errors, regulatory breaches, potentially exposed to fraud due to lack of visibility and control. In Mexico, the companies that import have the opportunity to work under the protection of the IMMEX program. The IMMEX program that is intended to promote the merchandise import industry; material or raw material temporarily and that within the country an industrial or service process is carried out to elaborate, transform or repair and to present export services, deferring the payment of General Import Tax, Value Added Tax and quotas compensatory (https://www.snice.gob.mx/cs/avi/snice/home.html). The IMMEX program is one of the many compliances that the company should be aware of, support, complete and follow up on in a timely manner and that, if it is not done, it would simply lose the benefit of the program. Moreover, having said this, even when the company has the program (IMMEX), the administration of Annex 24 has been quite deficient, and it entails the complexity of meet what is obligated: tax payments, it is not possible to have control of the customs duty, the return of merchandise, among others. Being a manufacturing company, it could not afford to stop for any reason, so is same critical level than downtime of equipment, a shortage of materials, but it seems that having its outdated customs control software was not a situation that alarmed the corporate .

On the other hand, but continuing with the long list of risks that this company ran; We can first highlight that, as is well known, foreign trade depends on several legal requirements and regulations. By not having the updated software, we already mentioned the risk of regulatory non-compliance, which could result in legal sanctions, fines and problems with customers due to the reputation that this would generate, since it is just as serious as if it were compared with tax evasion. In parallel, this company had a serious lack of control due to the lack of visibility, not being able to visualize the commercial operations in real time, it can reduce the monitoring of shipments and the opportunity to make strategic decisions in time would be lost. Not to mention that, to work in this way was inefficient and with many human errors: data loaded incorrectly or omissions of information. This leads us once again to put on the table the consequences that the company could face. Since in case make a mistake, omit any information or not be able to verify the origin of your merchandise, you could be subject to fines, and since our country is no exception, any company that manufactures and provides incomplete, false or incorrect information regarding merchandising, value and transport of goods.

What Are The Advantages To Obtain By Having And Keeping Your Customs Management System Updated?

Starting from the understanding of what customs management software is and what it is for, we must quickly deduce that it has the same purpose as any other software; which is to facilitate activities, operations and administration of your import and export transactions. In addition, the systems must be adapted to help reduce the costs, risks and delays related to compliance with customs operations, compliance that must be done before, during and after the transaction before the corresponding agency. Likewise, this helps you track your activities, and conclude your efforts. What is involved in Foreign Trade is the valuation of products and merchandise that move and transport across borders between countries, provide information on the country of origin of your imports and/or exports, comply with customs restrictions and controls, etc. Having software that helps you manage your foreign trade activities is also a great help in the control and standardization of the company's Supply Chain processes, ensuring that they are complied with and aligned with the most updated of the Customs Law standards ; import and export.

The importance of having and maintaining a Foreign Trade activity management software lies in understanding its scope. The advantages of knowing that it involves departments such as Finance; where obtaining the reports of each fiscal and accounting year are critical and dispensable, and that is usually look for to have a review for auditing issues of commercial invoices, or with the stamping on the Digital Fiscal Receipts by Internet (CFDI). Execute and receive the CFDI in an official file with the .XML extension, which is required in accordance with the regulations of the Tax Administration Service (SAT) and as a critical part it can always guarantee the payment in timely manner of the contributions declared by the executed customs transactions. For the Import/Export department, the advantage of having its trade administrator system updated is of great magnitude since this department is the one that manages and controls official foreign trade documents, such as: commercial invoices, certificates of origin, and contracts with suppliers., permits, petitions, etc.

Moreover, being able to share the electronic files of the suppliers makes collaboration easier, since they also mean that these files can serve as an alert if any document is missing. The organizing of documentation in electronic format is key to avoiding errors that, although they could be solved, could be avoided and waste the time of those who solve them; a clear example is in preventing payments of "Pedimentos" (*Custom Declaration*) with errors. This is also detectable in a good customs management system, not to mention core activities such as compliance with NAFTA (Free Trade Agreement) or NAFTA (North American Free Trade Agreement) requirements. As explained, in the company in question, the team members work under leadership of the Purchasing and Materials management so, the importance would be even greater when thinking about sustaining of the information of the customs management system. Since purchasing also has certain advantages; such as visibility in a simple and efficient way of expense invoices or the conceptualization and assurance of national invoices; CFDI. But as is known and unfortunately, it is frequent and well known that the areas involved do not report fixed assets, changes in Bills of Materials (BOM - Bills Of Materials), price adjustments, etc.

As an international company, it should be taken into account that the responsibilities before the Customs Law are of very high importance and their compliance can be achieved with software that helps you comply with them. From being registered in the Register of Importers and Exporters, with all the requirements that requires the SAT to carry out the activity and accreditation of the RFC "Registro Federal del Contribuyente" (Federal Taxpayer Registry). When having activity front of the register of importers and exporters, is required to work with a customs agency. The manufacturing company must be protected by its agent with an official and formal document declaring under protest to provide reliable information on the value of the merchandise front of the customs and being one of the most important responsibilities of all. One of the clearest advantages is the need to have a system that automatically controls inventories; this is an obligation in our country whenever the company imports merchandise under the temporary import regime for manufacturing, transformation, or repair in the "Maquila" program. Could not fail to mention that whoever keeps their foreign trade system up to date will have first-hand information and documentation for any argument with the effects of duties preferences, such as verification of the country of origin (COO; Country of Origin) and provenance of the goods

From the expert point of view, we could say that first, it must be understood that the administration of Annex 24 is mandatory for compliance before the law for any importer that is in the IMMEX program, and this cannot be better than using a software that manages robustly. This is also mentioned by Lic Rolando Martinez when briefly defining the importance of software as a tool for customs control:

The importance of any automated inventory control system for customs purposes, commonly known as "Annex 24", is that through them it is possible to comply with the legal provisions that are required of all importers, and especially those that have an IMMEX program authorized by the Ministry of Economy. For this, we must remember that IMMEX companies enter their inputs into the National Territory temporarily with the commitment to return them once they have been subjected to manufacturing, repair or transformation processes, and Annex 24 is the mechanism that allows demonstrating the return of temporarily imported merchandise and those that are still pending return, likewise, which allows the importer to prove the return of that merchandise through the reports generated in the system (Martinez, 2023)

We Must Update and Upgrade Our Software; What Does a Change of Customs Management Software Involve?

In the age we are living in, talking about telephone software, controllers, television applications, smart tablets, touch screens in cars, artificial intelligence, Industry 4.0 (fourth industrial revolution) have become routine topics of conversation, the evolution of computerized systems, autonomous systems, PLC automation systems, and many technological advances. Likewise, Foreign Trade software must advance in terms of updates, scope, ease of use and better quality deliverables. The company that owns and provides the service, or rather, the software, has restructured the system it offers and has an upgraded, improved and modern version. Supplier has notified its clients that the current system is planning to be EOL (End of Life) and that in a period of approximately six months with a possible extension to provide support only for fixes, but not for development or updating of any kind. The developers of the supplier company will no longer support their product. Likewise, the new Software is offered to replace the current one because it has been contemplated with new benefits. These are such as: the integration of data unified and integrated with the ERP of the company and external systems, these are the ones that maintained the integrity of the content of the management of special trade programs, legal reports available for compliance with Annex 24 and 30. In addition, security in the control of import and export activities before customs, absolute inventory control and an important part that was not available with the current system was the massive loading of Bills of Material: BOM. The supplier, acting in his role as negotiator and seller, provided the details of the scope of this new customs control package. On the part of the client, who in this case is the company that carries out the import and export activity, focused on making the comparison of the new software, reviewing each benefit and checking against the software that was currently available, validating licensing payments, support services, updates that the current system had but the new one did not require.

As part of the procedure that must be completed before a purchase, a comparison is made with RFQ (Request for Quotation) format, with the purpose to meet with the company's regulations. As the next step, it is necessary to search for a second customs control software provider, to make the comparison valid and to be able to have visibility of what both competing companies offered. A second supplier was found, to whom the current situation of the company was also clarified, and who, for obvious reasons, immediately agreed to present us with his RFQ as well. The details of your product and all the advantages that the company would therefore have when selecting them. With the complete information of both providers, a comparative table was created seeking to have a fair contest, the analysis of these two companies was made and the concepts to be analyzed were the same for both; Annual costs, each company presented its quote with this cost better known as: Payment of Licenses. The licensing is a framework contract that describes in detail what are the legal rights that the user of said software has authorized from the program, in other words, this would be a permission to use the software that is granted by the owner of the program to the company. As a second point lets mention one of the important parts that was: the review of the initial investment cost; The implementation of the project obviously requires resources, specialized personnel for the development and installation of the system, with its test environment, modifications, personnel that attends to customized requests, time, labor, etc. All this would be the cost of the implementation, which is an investment only at the beginning of the project. Payment can be deferred in stages. In the comparison, the delivery time is included; this type of project is develop in a period of between four to six months, depending on whether it integrates with the ERP that the company has, it should be noted that it cannot with any ERP can be linked.

In other words, the point of the data validation stage was also added; the difference between one system and another is which allows automatic validation and which will force you to do it manually. From the point of view of the systems structure that is required, the platform was reviewed. The interaction of the IT (Information Technology) department is critical and knowing the involvement in terms of security and control that will be given to the software to be acquired leads us to know if it will be supported by a Local Server or on the Cloud. And in the same way, the Software or Hardware that will be required and this can vary from Windows 2008, 2012, to any Windows Professional processor, using RAM (Random Access Memory) from 32 GB (Gigabyte) and 500 GB of space disk.

Among the most important things to mention in the comparison table is the connectivity with the ERP that the acquiring company has, knowing that current systems may not have this ability and the one that offers it will undoubtedly have an advantage over anyone else. The operational activities cannot be discarded and among the concepts also listed is the massive upload of Bills of Materials, the ease way with which it can be done and with which any type of report can be downloaded. Although it seems obvious, the revision of Annexes 24 and 30 had to be validated from the point of view of the introduction of information and its updating. This may or may not be limited and, like the aforementioned Annexes, the control of temporary imports for a company in the manufacturing industry that carries out import and export activities is essential. Also being a service, the acquisition of a license to use the Software requires support, for updates by the Customs Law in Mexico, by the CBP (Customs and Border Protection) of the United States, improvement of formats, reports, among others things. Review of the support provided by providers of this type is also warranted. The support can be Local, visiting you, attending to your needs personally, thru "tickets", and resolving remotely, other option is by recurring conferences or only when they are required. And without leaving behind the concepts that were required to review the number of users allowed. Although, we could say that in the Import/Export department it can be for one person who performs the operational functions. Therefore, the number of users allowed it may or may not be a limitation; if they offer you a single user, or five or no accounts limit, it may not be a critical factor, but it can be planned in the project review list.

The decision to select the new software supplier for foreign trade activities will depend on the needs of each company and, in this case, the decision was based on the connectivity that the software offered with the ERP that the company used, the platform on which the one that would operate the new system and the automatic data validation.

What are the Possible Problems We Face With a Change of Customs Control Software?

The selection of the Supplier of the new software was completed and, this was accomplished through a meeting with the company's corporate, where the representative team of the Import / Export area, personnel from the local and corporate IT area and the Plant Director participated mainly, with this, the arrangements, the plan and the implementation strategy started.

For someone new to a project of this size and importance, it may seem like too much bureaucracy, mistakenly thinking that it should be easy since it is a system. However, it is worth noting that a customs management system is the main tool for compliance with the country's government agencies; and that therefore must be protocol and legalize the project with a contract with which you will face the first delays. Said documentation must be reviewed by the legal area of the company, which if it belongs to the entity in Mexico. The communication will be easier and any question will flow quickly, but otherwise if the legal area is in another country, because it belongs to the corporate group. The questions will be many than if it were with someone in your own country and certain details must be explained that only the Project Managers of both parties will know and that are not stipulated or detailed in the contract. The legal area usually suggests changes; these changes are relevant since the legal department usually negotiates what has already been negotiated. Then, when the company's response and recommendations are sent to the supplier, the supplier will have the right of reply and will resend the contract with the notes on what was previously suggested to them. This will start a back and forth with the clause notes until both legal departments are satisfied and both companies are pleased with what was stipulated in the framework contract. Once the departments of the Legal area have agreed on the clauses, they have to be formalized with signatures.

The signatures can be from the legal representative, the lawyer representing the company or someone who has a higher hierarchy and with legal power over the company. Being large corporations, the people in charge of these types of firms are not always immediately available and it can take time to obtain them. This time can interfere with what has been determined in the project plan, and looking for the required signatures ends up being a task where the Project Manager is monitoring almost daily, reviewing and confirming that the documents have been signed. In addition, the counterpart on the supplier side has to do the same work once the signatures have been obtained, and now the baton will be on the supplier side until all the documents are signed by both parties; Project Scope (Statement Of Work ;SOW). In this is described in as much detail as possible what the project will include and what will not be included.

These are stipulated in clauses and will explain how the project will be completed, its phases and deliverables, costs (these can be completely broken down). Moreover, each stage and process of the project is detailed in a timeline. Termination Agreement; This clarifies the clauses with which the project must be governed in the event that for any reason either or both parties decide not to continue with the project once it has started, and the Order Form that can or not be mandatory. It will depend on whether the supplier requires it. In the Order Form is specified what the client is requiring, if we translate this into terms of purchase, the Order Form would be like a purchase order issued by the company for the purchase of the Software from the supplier.

Within the Scope of the Project; in the section where the times of the phases are specified, the dilemma of priorities in the company is entered. Only because the Project has a high priority, the sales cannot be left aside and for what the company is supposed to be profitable then, the timeline of the project begin to be questioned because the times do not align and you have to redefine priorities and/or the dates that are in the project phases. End users participate in some phases of the project, such as Design and Mapping, Training or user acceptance tests. For these stages, you will face the challenge of coordinating them, since the people involved in these stages must be available for up to three hours per day, with due time available and preferably without interruptions. Availability can bring with it a series of problems if it is not taken into account. If the users or the personnel that have been assigned to provide information or to attend a follow-up meeting are not available, the provider must agree to negotiate and reschedule the activity. In the event that there is a severe recurrence, the supplier has the capability to generate an invoice for the expense that originates from having to change the days and/or hours of the affected activities. On the other hand, it can be affected due lack of availability since by not being present or not paying due attention, in any instruction, direction or training, the personnel responsible for the tests or mapping may have failures and omit valuable information or requisitions for implementation development. By overlooking a significant change, it is very likely that during final tests or in the test environment the system will show errors. The problem is not detecting them, but having reviewed them and even that they could be avoided. Due to these omissions, the supplier and its developers must be notified again that modifications or details must be added to the deliverables, which by contract tend to cause extra charges. It being understood that the resources allocated to the project were calculated in a considerable time and having a developer redo what has already been worked on and tested will cause time invested and time delayed in the original planning of the project. Requests out of time are also specify in the scope of the project and is dictated, that if they delay the project they will have extra charges. The availability problem will not only be on the client's side, it can also be seen on the supplier's side. If the company that is acquiring the software is located in the north of the country, it may have a deferred schedule with the rest of the republic, leaving open a gap in working hours. Understanding that the hours of the operating personnel usually start and end earlier than the hours of the administrative personnel, this could even incur extra time for both parties if the times are not coordinated.

AVOIDING MISTAKES: WHAT ARE POSSIBLE PATHWAYS OF ACTION YOU CAN TAKE TO AVOID MISTAKES?

Getting it clear that the problems can vary, we cannot finish mentioning the various factors that can intervene in a foreign trade control software change project. Therefore, we also do not have a fully defined process, nor the recipe that will make the project have any setback, so we will take into account what we can do and that will be within the scope of our planning as a client. Based on the fact that, corporations have their legal area; this must be involved since the beginning, the legal representatives or whoever has the jurisdiction to sign the contracts and confirm the acceptance of the clauses. It is more than convenient to have internal meetings, that is, without the Supplier, so that the guideline is to always keep up to date and explain any doubts that you have regarding the legal part. It is also possible that the Project Scope document is not available from the beginning of the project, but the contractors or the software vendor usually has a draft. Being only a draft and not the formal and final document, it can be worked on in advance and thus you could eliminate the delay times that occur when back and fore with the questions of the clauses.

Likewise, with the draft you can gradually review with your legal department the issues that lead to a possible termination in the Termination Agreement document. With this strategy, you will be able to make as many revisions of the draft document as necessary. By having a draft, it also helps you to have the records of the changes that were required to your supplier in a managed and chronological way, for future explanations or simply to help for future projects. As a benefit, you will also be able to collect the signature of those responsible in a timelier manner by having all the information in advance and with the changes that were surely proposed in the draft by one of the legal parties.

Having the project in control implies, among many other things, communication, and to a certain extent, I would recommend "over-communication". This can help you spread the mission and vision of the project with all areas of the company. That, in turn, the leaders of these areas will be able to intervene with their mentions related to other projects in process, the priorities that should be considered, production planning, the times of the personnel involved, the directly operational export and import activities, and so on. Priorities will be defined one by one and teamwork will take place as a result of their own communication. However, if there were to be a manufacturing activity, a production order, a validation run with significant priority that has been directly or indirectly impacted, you will have the certainty that it was something that was determined jointly with all the area managers. Parallel to the definition of priorities, the times required for software users to be available will be exposed so, the meetings, activities and projects can be planned in advance. Could even be a viable option to train someone to can cover the activities while the meetings that we mentioned are held, which usually last up to three hours a day in various stages of the project. These activities of communication and planning will cause the personnel to have the availability to attend and will avoid delays and possible extra charges for the time consumption of the supplier's resources. As part of this strategy, we must not leave out the issue of the time zone difference due to the geographical area of the manufacturing company, which in this case is the client. The time zone difference should always be taken in consideration. As a suggestion, we can mention that a recurring meeting can be scheduled between the Project Managers of both companies, where they can be without time problems. At that meeting, they will be able to review as many pending issues as possible, and invite the other departments as required. Each activity or request will also be reviewed by both parties and must be challenged from the point of view of the schedule and agree in the event that any activity requires more time than normal business hours, and confirm whether or not this will cause extra charges.

Regarding the development of Software to meet the needs of the company, we must point out that end users have the greatest responsibility in everything related to process mapping, testing and special requests. Understanding by the special requests that could be special reports that include customized information for the company or information visible in the menus, bars or fields in the system. Therefore, end users can be prepared with a list that specifies, since the minimum details of the current process, common errors with current software, critical reports, with their respective periodicity and without generalizing any request. The more detailed the list, the better it can be defined with the developers. It is worth mentioning that this list of improvements and requirements for the system may have more than one topic that the developers will decline, but, in turn, you must have the expertise to always require alternatives to what you want to have, but for reasons of software, source code cannot be developed. In the end, the intention of having this list is that the scope of our needs is as robust as possible and to avoid making extemporaneous requests that after a certain stage of the project can delay the timeline and implicitly avoid an extra charge.

In terms of the operation per se, common problems or delays can always be addressed keeping in mind the potential consequences. In this case, is mentioned that during the transition there could be Custom Declarations or invoices not processed and this leads to a severe decision. Determining if this information or documents should be transferred to the new system or left in the current software. In addition, this can be defined as long as it is contemplated and fully understood what the consequence of doing so will be. Therefore, those who make the decision can be the leaders and managers involved, but those who have the details of how many, what amount or any other question is the personnel directly involved in the operation. And along this same line, we could determine that current software problems in general should be attacked in the same way. Firstly, detail the information, for example: if the problem is the Bill of Materials are not updated, you should start by reviewing: how many there are? How outdated are they? How long would it take to update them? How many resources do we have or do we need to do it? And this in order to avoid delays. If in the example there turn out to be enough Material Tickets to say that there were many, then the impact of bringing this problem to the new system would have to be reviewed. It will have to be defined if resources will be invested in the solution before or after the implementation of the new software. In short, it can be determined that no one will want to take problems with a system that expert users will be new to using. Users do not have enough practice to be able to review the other errors that are currently available. This could be a good reason to confirm that it could be the best option; fix the problem before the transition, but this decision could only benefit in the future despite the fact that the project and its planning would be at risk of not being completed on time. It is also recommended to list the number of problems in the current system and be able to work on their analysis with the necessary time and thus express analyzes and errors should be avoided.

LESSONS LEARNED

When a company decides to face the problems of its customs management system and chooses to update it with an improved version, it is necessary to understand the most critical milestones of a project of this magnitude. As is the history of transactions that you have and cannot be ignored. This is how Lic. Rolando Martinez mentions it, briefly explaining what should not be underestimated when making these implementations:

Any change in the system of Annex 24, must at all times keep the entire history of the operations (temporary imports, returns of materials, balances and bills of materials) that the previous system recorded, in order to be able to generate the reports of returns and current inventory that may be required by the Customs Authorities. The other option is to temporarily transfer the inventory of imported materials that is available at the time of migrating to the new software, leaving the history safe in the previous system to meet the information requirements of your customs operations. From the foregoing, it is important to highlight that all information that is transferred to a new software should not vary or generate new balances, as we remember that Annex 24 is the basis for generating the information on the export declarations and the discharges that must be present the companies certified in matters of SAT and IEPS. (Martinez, 2023)

Therefore, being immersed in the issues of Foreign Trade will give those involved in such a Project, the vision and confidence of what is deserved. They will have the understanding that a Customs Software is a living and dynamic system that always requires maintenance. In addition, any change in the Material Tickets, the currency exchange rate until the update in the rankings list. In addition, all this must always be considered in the first instance in the customs control system. It is of great value to understand the importance of good administration of Annex 24. That the customs law has clearly specified both the requirements to be met and the fines that apply in case of not doing it in a timely manner. Likewise, inventory movements must be monitored and supported by the MRP that the company uses.

Bureaucratic issues can not be thrown on deaf ears in the face of a project of this magnitude. The times that the project usually takes to collect the signatures and finalize the commercial contract agreements. The mission and scope of the project must be made clear since the beginning with all those involved, regardless of the amount of interaction they have with the software or if they are not from the technology department that must participate in the project. Always consider over-communicating with all staff to avoid unnecessary delays or charges.

The plans or activities that the company (client) will have within the time considered for the development and implementation of the project must be included in the project, as this will also help not to interfere with the times and priorities of the participating employees. Time planning in relation to priorities is essential in any strategic software implementation, regardless of its function.

The selection of a customs management software should be through a holistic analysis. It must allows seeing all the necessary points of view, the need of the company, current problems, future plans, the MRP system that is being used and its scope with this etc. In short, it cannot be a decision lightly and must be supported by a comparative analysis and approved by the senior management of the business. The system that is used, regardless of how modern and innovative it is, must generate the necessary diligence as long as its database, profiles and ranks are fed with reliable data.

To conclude, It can be say that the cost of customs management software will always be profitable compared to the consequences of not having it or having it and not keeping it updated. This comparison is unprecedented since the cost-benefit is obvious and the tool will provide the deliverables without any problem when analyzing the scope and results obtained.

The participation in a project of this nature such as the change of foreign trade control software is undoubtedly one of the most rewarding in my professional history. Since the scope of this drives you to understand from the protocol concepts of a framework contract or document, the Scope of the project and even the understanding of the platforms that usually support the software in the networks of a company's systems. Going deeper into the customs laws and their requirements is to be able to detect the meaning of each of these and know how much weight it has to maintain a good administration of almost daily transactions that a company in the IMMEX program must comply with.

REFERENCES

Blecker, R. A. (2014). The Mexican and US economies after twenty years of nafta. *International Journal of Political Economy*. https://www.jstor.org/stable/24696555

Martinez, R. (2023, May 26). *General Director of Agencia Aduanal Asesoría y Soluciones Aduaneras*. S.C. (P. Ramirez, Interviewer) Reynosa, Tamaulipas, Mexico.

Mendoza, J. E.-P. (2022). Comercio exterior y empleo en la frontera norte de México. (U. A. California, Ed.) Estudios Fronterizos, 23. doi:10.21670/ref.2207091

Servico Nacional de Informacion de Comercio Exterior. (n.d.). *IMMEX: Acerca de.* Servicio Nacional de Informacion de Comercio Exterior. https://www.snice.gob.mx/cs/avi/snice/progfom.immex.acercade.html#:~:text=El%20Programa%20IMMEX%2C%20es%20un,de%20servicios%20de%20exportaci%C3%B3n%2C%20difiriendo

Sánchez, R. B. (2021). *Home.* Diario Oficial de la Federacion. https://www.dof.gob.mx/nota_detalle.php?codigo=5639315&fecha=24/12/2021#gsc.tab=0

Tributaria, S. D. (2017). *Home.* SAT. https://www.sat.gob.mx/consulta/78455/exportadores

Tributaria, S. D. (2018). *Home.* SAT. https://www.sat.gob.mx/consulta/54336/importadores

Chapter 3
Transforming Next Generation-Based Artificial Intelligence for Software Development:
Current Status, Issues, Challenges, and Future Opportunities

Atharva Deshmukh
https://orcid.org/0000-0002-8039-3523
Department of Computer Engineering, Terna Engineering College, India

Disha Sunil Patil
Northeastern University, USA

J. S. Shyam Mohan
https://orcid.org/0000-0003-4521-3088
GITAM University, India

G. Balamurugan
https://orcid.org/0000-0002-5676-5235
Department of Computing Technologies SRM Institute of Science and Technology, Kattankulathur, India

Amit Kumar Tyagi
https://orcid.org/0000-0003-2657-8700
National Institute of Fashion Technology, New Delhi, India

DOI: 10.4018/978-1-6684-8088-5.ch003

ABSTRACT

All of the advances that we see in our lives are mostly based on software. Over the last several years, software development techniques have seen a significant transition. It simplifies and profits the firm, for example. Many customers' burdens are lightened as a result of their combined efforts. Today, software powers practically every area of the economy and is an integral aspect of a country's economy. Artificial intelligence is no exception. Businesses have embraced practically all developing technology in their software development. AI's effect on software development alters how businesses operate and makes software smarter. In this chapter, the authors have talked about the change AI would bring to software development and the role of could. The authors have also discussed traditional ways used. Towards the end of this paper, they have added open issues and challenges in detail.

1. INTRODUCTION

Conventional software development, which involves a number of steps such as preparing requirements, manually writing code, testing and designing software to ensure that the final product fulfils specifications, is not designed to accommodate modifications. Artificial intelligence (AI) is upending this process by enabling efficient and scalable practices that shorten time-to-market and boost productivity. Whereas several software firms have been in the early stages of AI implementation, the technology's adoption is progressively increasing across the spectrum. As per market research company Tractica, revenue from the use of AI technologies would reach $119 billion globally by 2025. It's important to note that artificial intelligence (AI) is not a substitute for human intelligence (HI). AI is constantly learning and delivering human interactions, all while saving time for software development teams by removing human mistakes in redundant operations. We are certain that AI with Software Development would strengthen development, agile test automation, automation-testing software, and the way RPA bots work with the help of the software as a result of this progress. Artificial intelligence (AI) tools are developed to ensure software development is more dependable, faster, and easier. The following are a few examples of AI's potential applications in software development:

- Requirement Gathering: As a conceptual stage of the SDLC (Software Development Life Cycle), requirement gathering necessitates the highest level of human involvement. Artificial Intelligence (AI) uses a variety of methodologies and technologies, such as Google ML Kit and Infosys Nia, to automate processes and reduce human participation to some level. Before moving on to design, this phase places a strong emphasis on the early detection of loopholes. Natural language processing is an AI approach that allows machines to comprehend the needs of the user via natural language then automatically generate high-level software models. As a matter of course, there are certain drawbacks to this strategy, such as the difficulty of balancing built systems. However, it is still a popular research issue today.

- Agile Project Management: The efficiency gains made possible by AI go beyond normal task management. The use of AI in software development can help engineers test, write, and deploy code to production teams in a more lean and agile manner. AI algorithms could be used for enhancing, cost estimates, volume estimations and project schedules by allowing development teams to prioritise code areas that need to be completed quickly and more precisely define potential failure probabilities. AI/ML algorithms dynamically examine massive code databases for anomalies, informing developers of possible next measures for protection. Glitches, Missing code alternate product or service names under the same code are examples of abnormalities. This isn't just useful in restoration, but data analysis can also help in erroneous pre-diagnosis.
- Automate Software Design: Since teams are distributed throughout the globe, designing software code is a time-consuming important, and, complex part of the development process. Developers, designers, Research and Development (R&D), and marketing teams must collaborate on a project by communicating efficiently and being transparent, which has been done manually until now. However, in the not-too-distant future, AI/ML algorithms could indeed help to streamline and automate the planning and design process by gathering data such as project stakeholders' location, names, products, business type and customer needs, to automatically create intuitive instructions upon which design approach to consider taking without involving manual intervention. It can really help programmers save effort, money and time, by automating the code design process.
- To present a definite answer, project planning and design necessitate specialist knowledge and experience. Designers face a difficult problem in deciding on the best design for each step. Retracts and a forward-looking investigation plan require dynamic design adjustments until the client achieves the desired result. By using AI technologies to automate some difficult tasks, the most capable approaches for project design can be used. Designers can, for example, employ AIDA (Artificial Intelligence Develop Assistant) to analyze the client's interests and aspirations and then use that information to design a suitable project. AIDA is a website-building tool that analyses numerous software design combinations and delivers the best-customised design based on the client's requirements.

- Automatic Code Generation: Getting a business idea and turning it into code for a major project takes effort and time. To address the issues of money and time, specialists have devised a strategy that involves writing code before beginning development. However, the strategy is ineffective when there are unknowns, such as what the targeted code is supposed to perform, because gathering these details takes significantly longer than building code from scratch. AI-assisted intelligence programming will help to alleviate some of the workloads. Consider how our system would grasp and translate our project idea into workable code if we explained it in our natural language. Though it may appear to be science fiction, artificial intelligence in software development can change the narrative! Natural language processing and AI tools will make this possible. Coding a complex, large project with several stakeholders, on the other hand, is often time-consuming and labour-intensive. AI coding assistance can significantly decrease development teams' labour while also enhancing productivity. Developers can increase their productivity by concentrating on more strategic and creative tasks, like enhancing the customer experience.
- Streamlining Software Testing: Testing is an essential component of any software development process. For development teams, detecting and preventing errors or bugs is a major challenge. Bugs and errors account for a considerable portion of software development costs. Early error detection necessitates continuous monitoring all throughout the project's development life cycle. Current software testing techniques, on the other hand, are inefficient, time-consuming and expensive, since many problems in the code are discovered after the application has been completed and given to end users/launched inside the marketplace. Trained AI and machine learning algorithms can assure error-free testing in much less time frame as compared to human testing, allowing code testers to focus on other critical duties like code maintenance. Development teams may test thousands or millions of lines of code with AI-enabled code testing prototypes. Case-specific tests can be handled by development teams, while time-consuming and repetitive tests can be handled by AI-assisted automation technologies. This leads to a reduction in errors since AI-assisted tests can scope and rectify faults with extreme precision, resulting in an overall improvement in software quality. In the near future, AI will be able to link with the cloud and do automated testing for software testers, fixing particular defects and delivering the product ahead of schedule, saving money, resources and resources time, while also generating a high return on investment for the company.

- Software testing is an important part of the software development process since it verifies the product's quality. AI and machine learning-based testing solutions include Functionalize, Testim.io and Appvance, to name a few. Incorporating AI tools enhances and improves coding, as well as makes error detection easier.
- Enhance Decision Making: Software developers spend a significant amount of money and time deciding which functionality to add to a product. By assessing the performance of existing apps and prioritising features and products for future development, AI can help speed up the decision-making process. Software firms can make data-driven strategic decisions rapidly as well as at scale, boosting impact on the market and improving profitability. Leveraging AI technologies such as advanced Machine Learning (ML), natural language processing, business logic and deep learning, software developers will be able to design better software faster. Machine learning technologies have the capability of learning from previous development projects and analysing the performance of current projects. AI in software development not just to facilitates the development and also produces more effective applications. Software developers can create better software with AI's strategic decision-making.
- Enhanced Data Security: During development, software security is a critical aspect that must not be overlooked. Data is collected by the system through network sensors along with the software installed on the customer's end. We can use AI to look at the data and use machine learning to discern abnormal behaviour from normal behaviour. Software Development in the Future Companies that incorporate AI within their development cycle can prevent issues such as incorrect notifications, alarms and delayed warnings.
- Improvement in estimate accuracy: AI provides an estimate software solution that incorporates examining historical data from previous corporate initiatives to uncover statistics and connections. It uses business rules and predictive analytics to provide accurate time, effort and cost estimates.
- The Future of Development: AI has enormous potential to reshape the future of software development. The availability of AI-enabled solutions allows software companies to deliver customer-driven experiences by providing application strategies according to business needs.

Software development teams may use AI algorithms and advanced analytics to make rapid decisions based on real-time data. AI applications conduct cognitive and complicated operations associated with human thinking, as opposed to robots that react to rules-based logic or offer pre-determined solutions. AI algorithms can automate the coding process by integrating data from a range of sources, such as sensors, remote inputs and microchips, to assist developers to write accurate code, resulting in more agile, scalable and efficient, processes. As a result, AI can be incorporated into software development to provide highly tailored products or services to customers. Artificial intelligence can have a massive effect on software development and design. Artificial intelligence's influence and potential benefits must be understood by software development organisations, not just in terms of how software is built, but also in terms of the program's nature. AI will play a critical role while designing, generation of code, and testing of software, among other things, and will be a game-changer in the near future.

In the next section Literature Survey. In section 3 of this chapter, we have discussed different software engineering methodologies like the agile method, different testing methods, etc. In section 4 we have discussed about different frameworks for AI development, ML life cycle and testing, debugging of AI/ML applications. Section 5 is about AI-based approaches to Software Engineering. The next section 6 is about the use of the cloud. Section 7 is about different technologies of Software development. In sections 8 and 9 we have discussed about Open issues and future research opportunities. Section 10 is about the change AI will bring to software development and the last section 11 conclude this works in brief.

2. LITERATURE SURVEY

Despite the fact that Artificial Intelligence (AI) is becoming a buzzword in relation to self-organized IT technologies, its implementation in software engineering has received little attention. The analysis concludes that a) the automation of long and tedious routine jobs in software testing and development utilising algorithms, such as documentation and debugging, b) structured and systematic analyzing large data pools to identify and analyze patterns and novel information agglomerations, and c) the comprehensive and systematic assessment of this information in neural network models are major achievements and future potentials of AI. As a result, AI aids in the acceleration of development processes, as well as the reduction of development costs and increased efficiency.

Artificial intelligence, identified as machine-based reasoning, interpretation and perception, of mental and environmental constructs (Kakatkar et al., 2020; Poole & Mackworth, 2010), has the potential to multiply software engineering strategies and approaches, such as the methodical methods to do design, assessment, implementation, software analysis, reengineering, test, assessment, and maintenance (Laplante, 2000). For example, Jarrahi (2018) sees AI's key capabilities of software engineering in relation to language processing (i.e., recognition and interpretation of human language), machine learning (adoption and workflow analysis), as well as machine vision (target and strategic concentrated machine reasoning and problem). Muenchaisri et al. (2019) similarly discuss AI functions such as neural networks, natural language processing and machine learning.

Based on more of a quantitative and systematic categorization of previous studies, Savchenko et al. (2019) review of 54 studies helps to identify the fields of IoT and big data technologies, machine learning, system analytical techniques, design and programming assistance tools, and knowledge management and recommendation tools. Although function-related classifiers of AI within software engineering are experimentally supported, there is a danger that unexpected domains that AI may plausibly serve will be overlooked because they have not yet been empirically studied. In a variety of case studies, Andreou and Stylianou (2016) analyse the optimization outcomes of numerous algorithms in multi-objective assignment optimization involving duration and cost objectives and find AI-based approaches superior to traditional linear planning.

Zhang et al. (2016) employ a target-driven technology-road-mapping decomposition model that depends on an adept understanding of the database and uses fuzzy and semantic group analytics methodologies to assess project perspectives dependent on several interdependent determiners. The client and developer verify the software product's functioning in practical, analyse and detect faults, and modify the product to practise needs during the testing phase (Seffah et al., 2005). To aid software integration and testing, AI employs machine learning and pattern recognition methodologies (Meinke & Bennaceur, 2018). Filieri et al. (2015) developed a runtime decision algorithm-based engine that allows an application to adapt to changing circumstances. Depending on network needs, the self-adaptive system continually reconfigures software components.

This routine reduces the need for human assistance and updates while still ensuring on-time security, programme stability and adaptivity, AI also aids software modernization by utilising machine learning and pattern recognition methods. Due to insufficient documentation, structural information is commonly lost in ancient codes. To extract coherent collections of code, AI pattern recognition algorithms are useful. To trace and check their functionality, machine learning functions are applied (Alhusain et al., 2013). Pattern tracing functions eliminate unnecessary code pieces and generate implementation artefacts and test software functions automatically (Van Hoorn, et al., 2011). Fenton et al. (2010) propose a Bayesian network approach for optimising cost and quality outcomes at the same time. Bayesian models, unlike traditional optimization models, can incorporate a high number of co-determiners and coefficients, as well as deal with missing and ambiguous data. In order to achieve an ideal work assignment and maximise team performance, Athavale et al. (2013) employ AI to predict the interconnections between human beings and their environment in software project operations. To assemble performing teams as measured by output quantity and development speed, the model takes into account human personality traits and affective states, as well as competencies, learnability, and individual interactions. The stages of software development are referred to in process-related classifications of AI in software engineering, which examine how well AI can support software engineering at each stage.

This method is likely to avoid the bias of ignoring AI development requirements and shortcomings, which are common in functional classifications. Only two evaluations have been found that use process-related classifications: Sorte et al. (2015) and Padmanaban et al. (2019) both recommend categorising Ai systems into three stages of the software engineering life span: software testing, creation, maintenance and deployment. A six-stage model of the software engineering life cycle is used to implement the review, which has found widespread use in academic software engineering literature (Leau et al., 2012) and software development practice (Ramadan & Widyani, 2013; Velmourougan et al., 2014). Project planning, problem analysis, software design, implementation in software code, software testing and integration, software maintenance and support are all included in this step.

3. SOFTWARE ENGINEERING PROCESSES AND METHODOLOGIES: IN GENERAL

The software engineering process entails tasks such as maintenance, testing, coding, design, analysis, and requirement gathering to manage the development of software. Different approaches to the delivery of software and software development are referred to as software engineering methodologies. The software process model is a simplified version of the process being discussed. A simplified model of the software process may also be defined. Each model depicts a process from a unique viewpoint. Different types of software process models can be implemented using basic software process models.

3.1 Agile Methods

Agile is a set of methods that a team uses for managing a project by breaking it into smaller phases and regularly collaborating with clients. Every level of the software development for the project is regularly monitored. The benefits of agile constitute the fact that, unlike traditional waterfall methodologies, both testing and development processes are synchronised and concurrent. On the market, there are a variety of agile approaches to satisfy the needs of any project. Regardless of the fact that there are numerous agile approaches, they are all based on the main ideas of the agile manifesto. As a consequence, any framework or behaviour that follows such principles is described as Agile, and despite the various agile methodologies used by a team, the benefits of agile methodology can only be completely realised with the engagement of all involved parties. A dynamic approach is required when selecting the appropriate agile methodology from the several forms of agile methodology. The advantages and disadvantages of agile methodology must be carefully assessed when choosing the framework for one's organisation in order to attract talent and offer excellent digital experiences in this fiercely competitive sector.

3.2 Business Process Design and Deployment

The business process is a set of procedures that allow a company's activities to run smoothly. These workflows are made up of a set of actions that must be executed in a specified order in order to accomplish objectives. The Business Process Design (BPD) is the process of creating new processes from the ground up in order to meet a company's objectives. The purpose of BPD is to create scalable and easy-to-replicate processes and workflows. BPD helps us to avoid superfluous stages in order to streamline a process and save resources by ensuring that each team member understands the exact duties and order of action. Deployment checklists are developed, delegated, coordinated, updated, executed, and reported on by Process Owners. Because Procurement, Information Technology, Training and Development, and other departments are required to establish the proper environment that supports the process, the checklist spans across businesses.

3.3 Program Specification and Modeling Languages

A programme specification outlines the intended outcomes of a programme; its primary goal increases understanding rather than implement it. The programming technique is built on the foundation of specifications. A specification is indeed a technical agreement between a customer and a programmer or company that establishes a shared understanding of the software. A client utilises the specifications to guide the usage of software, whereas a programmer utilizes it to drive the creation of a required programme. There are sub specifications also, each specifying a sub-component of a particular programme, which might be generated from a complicated specification. The building of such sub-components would subsequently be assigned to other programmers, allowing a programmer to serve a client.

Any textual or graphical computer language that specifies the design and creation of models and structures according to a set of frameworks or rules is referred to as a modelling language. The modelling language is a subset of the artificial language and is comparable to it. The modelling languages are mostly used in software development to create models for new software and systems. Modelling languages are largely graphical and textual in nature, although they are divided into 4 groups depending on the specification and requirements of a domain in use:

- Language for data modelling
- Language for system modelling
- Languages for object modelling
- Modelling language for virtual reality

The Unified Modeling Language (UML) is a popular modelling language for creating visual representations of systems and objects.

3.4 System Validation and Verification

Verification is a process of establishing whether a product meets the specifications or standards that have been set forth. Validation is the process of determining if a delivered or planned system will fulfil the sponsor's operating requirements in the most realistic environment possible.

- The validation phase examines and validates the real result whereas the verification process examines documents, designs, codes, and programmes.
- Validation requires the execution of code, whereas verification does not.
- Validation determines if software satisfies needs and expectations, whereas verification determines whether software verifies a specification.
- Validation determines if software satisfies needs and expectations, whereas verification determines whether software confirms a specification.
- When comparing Verification versus Validation testing, the validation phase occurs after the verification step.

3.5 Program Analysis Tools

The Program Analysis Tool is an automated tool that takes the source code or executable code of a programme as input and produces observations about the program's features as output. It describes the program's size, complexity, commenting quality, adherence to programming standards, and many other features (refer to Figure 1).

3.6 Software Mining Specifications

Software mining is a type of knowledge discovery that entails analysing existing software artefacts in the context of software modernisation. Reverse engineering is a term used to describe this procedure. Typically, existing programme information

Figure 1. Classification of program analysis tools

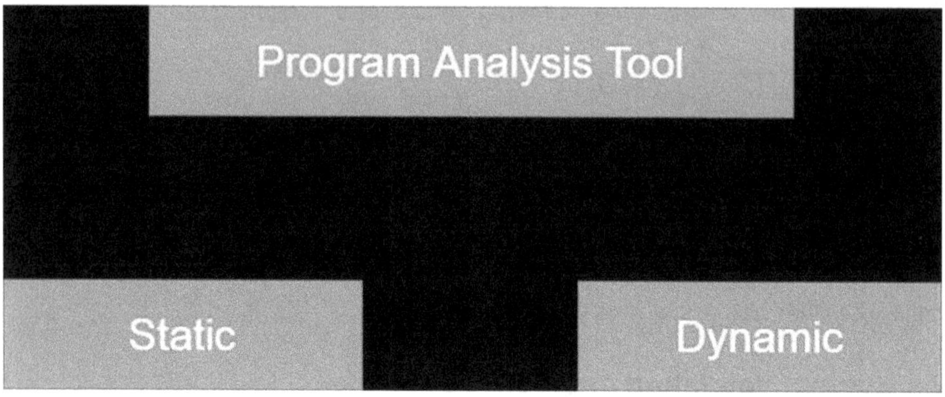

is offered in the form of models, to which particular queries may be performed as needed. A common approach for expressing knowledge gathered from existing software is an entity relationship. Because existing software artefacts hold tremendous economic value and are critical for the growth of software systems, software mining is closely connected to data mining. The structure, behaviour, and data handled by the software system are all addressed in knowledge discovery from software systems.

3.7 Software Testing

Software testing is a technique for determining if the actual software product meets the expected criteria and ensuring that it is defect-free. It entails the use of manual or automated methods to assess one or more attributes of interest by executing software/system components. In contrast to real requirements, software testing's goal is to find mistakes, gaps, and missing requirements. Program testing is important as it allows any faults or mistakes in the software to be found early and fixed before the software product is delivered. A well-tested software product provides dependability, security, and excellent performance, which saves time, money, and improves customer satisfaction.

3.8 Engineering for Quality Requirements

The application of quality engineering methods to the SDLC (software development lifecycle) is known as software quality engineering (SQE). Instead of being layered on top of current workflows, software quality engineering is tightly linked with existing DevOps and agile processes. This is similar to the DevSecOps team's shifted-left testing methods and is meant to discover concerns early on. Engineers play a crucial role in software quality assurance. Quality engineers create, execute, and manage rules and procedures to guarantee that development processes are of high quality. This necessitates a deep grasp of current evaluation methods and technology, such as big data analytics, automation, and AI. The primary goals of software quality engineering are:

- Process control and oversight
- Implementing standards and metrics
- Data collection and analysis
- Test development
- Identification of issues and solutions
- A follow-up to ensure corrective actions

3.9 Model-Driven Development

Model-Driven Development (MDD), is a software development process that allows users to create complicated systems by abstracting pre-built components into simpler abstractions. Rather than explaining, all these visual building pieces demonstrate the business's needs and potential solutions to technical issues. Model-Driven Development is the most crucial principle of low code development it is also the link that connects business domain specialists with IT, allowing them to cooperate and turn ideas into useful solutions.

3.10 Software Economics and Metrics

Software Economics in Software Engineering is a mature academic topic that normally works with the most challenging and difficult challenges and concerns of valuing software and evaluating or estimating the costs typically associated with its creation. These issues and challenges are outlined by Boehm and Sullivan (2000), who also show how software economics ideas may be used to better software design and development.

3.11 System Modeling and Simulation

Modelling is the process of expressing a model, including its design and operation, in the field of modelling and simulation. This model is based on a real-world system, and it aids the analyst in predicting the impact of system modifications. The operation of a model in terms of time or space, which helps examine the performance of an existing or proposed system, is known as a simulation of a system. We will explore the concept and categorization of Simulation and Modelling, as well as its architecture, application areas, and other major concepts, in this lesson.

3.12 Traceability of Software Artifacts

Software artefact traceability is widely acknowledged as a critical component of successful software development and maintenance. "The capacity to follow the life of a requirement in a forward and backward manner" is how software artefact traceability is defined. Traceability between software artefacts can be useful for a variety of purposes (Oliveto et al., 2007).

3.13 End-User Development

End-user development is when a software program's end-user is in charge of creating new apps or assets. This is in contrast to more fundamental software development, in which the business responsible for generating a program also develops all of the program's applications and assets. The fact that such development may speed up development time and make users feel more connected with the program are two major benefits. End-user development has been used in a variety of applications, including video games, graphics programs, and scientific modelling tools.

4. SOFTWARE ENGINEERING AND ARTIFICIAL INTELLIGENCE/MACHINE LEARNING

From robots to Google Siri and now the new Google Duplex, artificial intelligence appears to be making significant progress toward becoming more humanistic. Machine learning and artificial intelligence are in high demand. As a result, the community has grown, which has resulted in the development of various AI frameworks that make learning AI simpler. Artificial Intelligence (AI) is the computing industry's future. And, as the need for AI grows, a growing number of programmers are becoming familiar with this technology. Information with respect to software engineering for AI can be discussed below:

4.1 Development Framework for AI Applications

The demand for machine learning and AI has grown exponentially. Additionally, the community itself has increased as a result, and that has led to the evolution of some AI frameworks that make learning AI much easier. Now, few of the simulations/ framework will be discussed here as:

- Caffe: Convolutional Architecture for Fast Feature Embedding is an acronym for Caffe. It has a strong design that optimises without hard coding and follows configuration-defined mechanisms. This may also be used to switch between GPU and CPU. Caffe is great for industrial and scientific tasks since it can process over 60 million pictures per day on a single NVIDIA GPU. MATLAB, C++, Python, and CUDA to command-line interfaces are all supported by the AI framework. It's quite easy to use Caffe to create a CNN (coevolutionary neural network) to recognise a photo.
- Torch: The torch is a scientific computation system that can do both scientific and numerical computations. It creates algorithms that are quick, versatile, and easy to use. The torch, which is a Tensor Library akin to NumPy, looks to prioritise GPUs. This is included in LuaJIT and provides a basic C/CUDA connection. Having a large number of algorithms, this improved performance and encouraged deep learning analysis. Torch consumers come with simple libraries, enabling the modular implementation of networked artificial logic systems. This improves with processes like cutting and indexing thanks to a flexible N-dimensional array. It also incorporates neural networks and linear algebra methods.
- Scikit-learn: Scikit-learn is a commercially licensed AI framework and one of the more accessible AI methodologies. That's a Python software that supports unsupervised ML and supervised ML. It is a flexible AI generation approach that supports grouping, dimensional reductions, clustering algorithms, and regression, as well as model collecting and pre-processing.

4.2 Machine Learning Lifecycle Management

A development lifecycle that enables learning models for generating bespoke Machine learning and apps has become extremely important as machine learning gains popularity in enterprises. As a result, it's critical for data-driven businesses to select a Machine Learning platform that interacts with different Machine Learning frameworks. Machine Intelligence (MI) Life Cycle Management is significant as it defines the role of each and every employee in an organisation, from business to engineering, in data science activities. Although there are a variety of established technologies that supports every phase of this lifecycle, also there are few solutions that connect those components all together into a unified machine learning platform. To support a model's lifespan, one must be able to manage the numerous artefacts related to ML and their relationships, as well as automate deployment. For deployment, visualizing, versioning, and storing of artefacts, a lifecycle management service created for this purpose should be used.

4.3 Testing, Correctness, Debugging, and Interpretability of AI/ML Applications

AI/ML short for artificial intelligence (AI) and machine learning (ML)—represent a significant advancement in computer science and data processing that is rapidly revolutionising a wide range of sectors. Businesses and other organisations are faced with a growing tsunami of data that is both extremely valuable and increasingly difficult to acquire, handle, and analyse as they undertake digital transformation. To handle the massive amounts of data being collected, mine it for insights, and act on those insights after they've been discovered, new tools and processes are required.

- Testing: For ML models, white box and black box testing are employed. Generating training data sets that are large and comprehensive enough to meet the goals of ML testing is a substantial challenge. During the development phase, data scientists compare the model outputs expected results to the actual results in order to evaluate the performance of the model. Some of the methodologies for black-box testing ML models are Model performance testing, Metamorphic testing, Algorithm Ensemble. Backtesting is the process of putting a prediction model to the test using past data. This approach is commonly used in the financial industry to measure the performance of prior models, especially in credit risk assessments, fraud detection, trading, and investing.
- Correctness: The number of classifications a model successfully predicts divided by the total number of predictions produced is known as model accuracy. Model quality always relates to fit, not accuracy, in the absence of standards. We don't dismiss the entire model if we disagree with a single prediction; rather, we embrace the reality that some forecasts will be incorrect and strive to quantify how often a model is incorrect given the facts in our case. In other words, an evaluation will decide if a model is appropriate for a certain problem and beneficial in solving it. This is a big departure from traditional software testing and is more akin to software validation tasks such as acceptability and user testing.
- Debugging: Model debugging is a new field that focuses on identifying and resolving issues in machine learning systems. Model software testing and risk management are all now used, in addition to newer developments. Model debugging tries to test Machine learning models like code and investigate complicated response functions of the ML model and make decisions to limits, find and fix errors in ML systems to improve fairness, security, accuracy, and other issues.

- Interpretability: The more interpretable an ML model is, the easier it is to understand why particular judgments or predictions were made. If a model's judgments are simpler for a person to understand than the decisions of another model, the model is more interpretable. Using only a subset of algorithms that build interpretable models is the simplest technique to obtain interpretability. Interpretable models such as logistic regression, linear regression and decision tree are frequently utilised.

5. SOFTWARE DEPLOYMENT AND OPERATIONS

The use of AI-based approaches to Software Engineering challenges has recently sparked a spike of interest. Recent developments in Search Based Software Engineering, as well as long-standing work in machine learning and Probabilistic reasoning for Software Engineering, are examples of the work.

5.1 DevOps

Advanced technologies such as ML and AI address a variety of problems and reduce DevOps' operational difficulties, allowing companies to shift quickly. The different ways in which AI is altering DevOps are listed below.

- Enhanced Data Access: In DevOps, a lot of data is created every day, and the team is having trouble accessing it. Artificial Intelligence can assist gather data from many sources and arranging it. This information will aid in the analysis and provide a clear picture of current trends.
- Security: Distributed Denial of Service (DDoS) is a popular attack these days. It may be used against any company or website, be it small or large. ML and AI can aid in the detection and management of these attacks. An algorithm can be used to distinguish between abnormal and normal circumstances and then take appropriate action. To improve security, DevSecOps may be enhanced with Artificial Intelligence. For identifying abnormalities and threats, it incorporates a centralised logging architecture.
- Software Testing: AI aids in the creation of processes and the testing of software. Functional testing, Regression testing, and user acceptability testing are all forms of testing used in DevOps. These tests generate a significant amount of data. AI recognises the pattern in the data and then determines the coding techniques that resulted in the problem. As a result, the DevOps team will be able to use this data to improve their productivity going forward.

5.2 Container Based Software Artifacts

Developing cloud-based AI systems necessitates the incorporation of these capabilities into containerized microservices. This entails incorporating Artificial Intelligence and other application logic into Docker images that could be managed via Kubernetes or other could based orchestration backbones utilising Java, python, and other languages. To create efficient AI microservices, developers must break down the core application capabilities into modular pieces which could be deployed in cloud-native settings with low resource constraints. In a cloud services context, Artificial Intelligence microservices are dynamically containerized and orchestrated inside lightweight interoperability fabrics.

5.3 Handling Container Image Sprawl

Container sprawl refers to the practice of accumulating an excessive number of containers. There are some distinctions between running a physical data centre and a cloud native container system, the main issue is the same that is cost. Spinning up a large number of containers produces business concerns, such as cloud computing expenses and administrative issues, which can lead to inefficiencies in most cases. To be clear, while deploying containers is far more convenient rather than establishing a new physical server or servers, the costs associated with doing so can easily spiral out of hand. The best approach to utilize the most out of the capabilities that containers provide while keeping expenses under control is to implement appropriate virtual machine and container policies.

5.4 Green Data-Center Management

A greener data centre will save power costs, increase cooling capacity, and improve system dependability while lowering the ownership cost considerably. Energy consumption is one of a data centre's biggest costs, as per the Storage Networking Industry Association. It's also one of the top ten concerns for data centre operators. Adopting a green strategy for data centre energy and environmental management might be costly. These data centres, on the other hand, provide a variety of benefits over time. The following are some of the advantages of green data centres: reduces $CO2$ emissions, reduces the use of electricity, stresses the usage of renewable and environmentally friendly data centres, reduces water usage.

5.5 Fault Isolation and Repair

A fault is an incorrect step in any workflow or data specification in a computer program that is accountable for any program's unwanted behaviour. Errors can be caused by hardware or software flaws or bugs. An error could be defined as a component of a system that causes the system to fail. A program error is essentially a signal that a failure has occurred or is about to occur.

6. SOFTWARE FOR THE CLOUD AND CYBER-PHYSICAL SYSTEM

A key component of critical infrastructure, cyber-physical systems are so important to the country that their destruction or incapacity would have a crippling effect on social stability, public health, economic security, national security, or any combination of these factors (Gao et al., 2015).

6.1 Security

Cyber-physical systems are intelligent, a large network of physical devices (connected to web), and standardised but it also introduces a slew of new security issues. About 90 percent of attacks can be detected using signature-based approaches. Traditional procedures can be replaced with AI to enhance detection rates by up to 95 percent, but there will be an explosion of false-positive. Combining traditional approaches with AI would be the greatest approach. This can lead to fewer false positives and we can also achieve a 100 percent detection rate. By incorporating behavioural analysis, organizations may employ Artificial Intelligence to improve the threat hunting process. For example, by analysing large amounts of endpoint data, AI models may be used to generate profiles of each and every application inside a company's network.

6.2 Data Obfuscation

Data obfuscation is the practice of masking sensitive data with data that seems to be true production data, rendering it worthless to malevolent actors. It's mostly utilised in test and development environments, where developers and testers want accurate data to design and test software but don't need to view the real thing.

6.3 Application Driven Resource Allocations

UDNs which stands for Ultradense networks have emerged as a potential architecture for future data traffic support, however, this deployment scheme would present significant issues in Application Driven Resource Allocations. UDN situations can benefit from artificial intelligence, which allows intelligent communication devices to do resource allocation. The concept of event-triggered Q-learning was brought in the study of power and subchannel distribution in Ultradense networks in this work (Zhang et al., 2019), and an event-triggered Q-learning-based resource allocation method was suggested by the authors.

6.4 Cross-Cloud Deployment

A multi-cloud approach can provide us with this extra degree of flexibility in a number of ways. Running our AI projects across several cloud service providers adds a layer of stability, decreases the chance of downtime, and gives companies business continuity according to the use-cases demand. While the idea is simple to grasp, putting it into practice is a little more difficult. Each cloud provider has built its own data management system, making data transfer across clouds challenging. A cross-cloud strategy overcomes this problem by properly managing data with each provider while offering a consistent end-user experience, regardless of where the data and apps are stored. Multi-cloud is also critical for maintaining continuity of operations in the case of a disaster (Horn et al., 2019).

6.5 Performance Engineering for Cloud Apps/Cyber Physical System

Since the advent of computer programming, software engineers have been tripping over false assumptions. Take, for example, a ticketing system for airlines. Airlines had little or no online services ten years ago. Only if we were using the appropriate browser we could search for flights and buy tickets.

6.6 Self-Adaptive Systems

Self-adaptive systems are intended to give formal assurances concerning the robustness and efficacy of adaptation mechanisms. When it comes to dynamic adaptation, the computing efforts that are required to establish assurances impose significant limits. For addressing these issues, a hybrid strategy that integrates AI, control theory, and software engineering to design for software self-adaptation is recommended. A performance-tuning dynamic and hierarchical system manager. A hierarchically constructed components design seeks the separation of concerns to a dynamic solution due to the gap between high level requirements definition and the internal knob behaviour of the controlled system (Caldas et al., 2020).

7. OTHER USEFUL COMPONENTS/ TECHNOLOGIES FOR SOFTWARE DEVELOPMENT

Software development is the process of creating computer software using one or more specialised programming languages in order to meet certain corporate or personal goals. Software development is typically a well-thought-out process involving a series of processes and stages that culminate in the creation of functional software.

7.1 Software Engineering Education

Several research investigations in the United States found that the government relies on poorly understood and highly massive software systems, which are prone to frequent unanticipated breakdowns. Several studies have also revealed that many software projects were cancelled due to being costing and late more than expected. A software failure can lead to a company failure. Because most company applications are shifting to an e-business atmosphere, software quality is becoming increasingly important. These environments are becoming more complicated and sophisticated as a result of modern techniques and the growing role of software in daily life. In this regard, it is the responsibility of Software Engineering (SE) professionals to ensure the quality of SE education (SEE). As a result, it is SEE's obligation to equip SE professionals by providing them with skills to match the software industry's requirements. According to studies, there is a significant gap between software industry needs and education for future managers to adequately manage this complicated environment. We recognise that software engineering is the speediest evolving engineering application of engineering and that the majority of software development organisations' jobs are diversified. It is also critical that graduate students gain extensive experience in a variety of fields. If the pupils are well-versed in emerging technology, this will be a plus. It also suggests that communication, skills in continuing education, professionalism, multidisciplinary abilities and practical experience are all vital aspects of software engineering education. These elements are not expressly addressed in existing guidelines and studies. While there may be other aspects to consider in software engineering training, we consider the ones we've mentioned are critical since software engineering evolves at a faster rate than any other engineering discipline.

7.2 Deep Learning

Due to the success of deep learning in pattern recognition and data mining, academic researchers and industrial practitioners have been increasingly incorporating deep learning into SE problems in recent years. Deep learning assists SE participants in extracting requirements from predicting software flaws, generating source code, and natural language text, among other tasks. SE can be separated into five phases, as proposed by traditional SE models such as the Incremental Model and the Waterfall Model. These phases include software design, maintenance. requirement analysis, development, and testing. With software design, software design patterns can be identified, such as synthesis and code recommendation, programme learning, and so on. Additionally, software testing and maintenance are important steps to consider while attempting deep learning. Reliability estimation and Defect prediction, as well as defect prediction, malware detection, changeability estimation and reliability, and effort estimation or development cost, are all included. People with the title software engineer-deep learning are responsible for activities such as modelling, data engineering, AI infrastructure and deployment. Define data needs, label, move data, examine, augment, clean, and gather are all data engineering subtasks. Training deep learning models, reading research papers, setting evaluation criteria, and searching hyperparameters, are examples of modelling subtasks. Subtasks for deployment include converting prototyped code to production code, setting up a cloud infrastructure to deploy the model, and optimising response times and bandwidth usage. Maintaining reliable and Building, fast, scalable and secure, software systems to assist employees working in modelling, data engineering, business and deployment, are examples of AI infrastructure subtasks. The deep learning method is learning for learning representations of data with several degrees of abstraction using computational models built of numerous processing layers. For example, we use the deep learning method AutoEncoder to summarise bug reports, which is a common SE task. Bug reports are writings that describe software flaws. Bug report summarising seeks to construct a summary by extracting and highlighting relevant sentences from a bug report to reduce reading time when faced with a large number of them. Researchers use AutoEncoder to unsupervised encode the words in bug report sentences in order to identify informative sentences.

7.3 Robotics

Robotics and software will have to fulfil the economy's ever-increasing demands and facilitate production in the next years. To be more specific, here are the five developments to look out for in the domain in question:

a) Robotic systems will be more user-friendly: The complex industrial robotic system of today necessitates highly skilled specialists, which is one of the main roadblocks to the widespread use of robotics, aside from the high cost. As a result, in order to democratise the usage of robots in the future and make them accessible to the general public, we will require increasingly complex software to handle robots on a daily basis. Is it possible to put this into action right now? DMG MORI, a German manufacturer, is demonstrating how this would be achievable. Their personnel use straightforward mobile apps to control robotic solutions, allowing low-tech specialists to programme the automated systems.

b) More customisation to satisfy specific needs: Building highly specialised robots is currently an expensive pastime. Robotics development, maintenance, and use will all be cheaper if more adaptable solutions are developed. Furthermore, the customizability of robotics must match the unforeseeable wants of the future market in order to establish a vast path in the future.

c) Improved management for improved manufacturing control: Better administration of fully automated technology reduces costs and saves time. Though we can't expect robotics to be error-free, we can demand shorter incident reporting and resolution times. Furthermore, better management is required to improve strategy-making and prediction. This is due to the need for stability, which is a must-have characteristic for software developers.

d) Traceability for greater feedback: Knowing where, when, and how our equipment is working is critical in the organisation. As a result, we may rapidly identify overabundant robotic labour in one operation chain and redistribute it to wherever it is most needed.

e) Modification: To introduce upgrades to robotic system design is the driving force of life, according to evolutionary theory. Things may change faster than they do now in various sectors of robotic implementation. As a result, the most difficult challenge for software developers will be fast updating automated equipment and upgrading it as needed.

f) Software quality: Whether in the field of enterprise robots or self-driving vehicles, the efficacy of our solution is determined by the professionalism of the developers and software quality. As a result, mobile app development businesses and software like MLSDev are now following a popular trend: engaging in the professional development, training, and mental well-being of those responsible for the digital solutions of the future.

7.4 Natural Language Processing (NLP)

The fields of natural language and software engineering processing are intertwined. They are both engineering fields and computer science. Natural language processing is a multi-computer method that works with natural languages. Natural language processing is a branch of artificial intelligence, computational linguistics and computer science, concerned with the interactions of human (natural) languages and computers, and in particular with programming computers to process huge natural language corpora efficiently. Analysing and perceiving visual scenes, speaking natural and understanding language such as computing arithmetic, logic operations and English, having memory, decision making looking forward to and inferring for generalisation and learning from new experiences, are all examples of human natural intelligence.

The digital transition, the Internet of Things, and the use of social networking sites have all resulted in a large amount of semi-structured or unstructured text data. With zetta bytes of data in 2020, this age is dubbed the “big data” era. Text accounts for up 8% of all data on the internet. Thus, in the big data era, AI products and services that recover information from AI are more important in structured numerical computing. The most recent natural language processor, GPT-3, was trained using the whole internet text data and was launched in June 2020. The history of NLP and the present situation of the GPT-3 NLP model are discussed. There is a progression of programming languages between the first and fifth generation, as well as shifting programming trends.

Text mining has recently been developed to teach a computer to comprehend human communication. Natural language is a set of terminology that abides by text mining that has just emerged to teach a machine to understand human communication. Computer Vision is a term that refers to the study of Many common methods, approaches and tools, that can be employed in the creation of NLP software, according to Software Engineering. There are numerous prospects for the application of NLP to improve SE theory and practice in the context of Software Engineering (SE). Many studies have recently been conducted to determine the extent to which large code corpora retrieved from Stack Overflow, GitHub, and other sources can be analysed using statistical NLP models and algorithms, allowing revolutionary advances in translation, speech recognition, comprehension, and other areas to be applied in SE.

We have managed NLP in SE under the following headings: a) NLP in Umbrella Activities b) NLP in the Software Development Life Cycle.

The writing survey provided insight into how Natural Language Processing can be applied to SE improvements. It was difficult to conduct a Systematic Literature Review (SLR) due to various shortcomings in obtaining various false positives. The common dialect handling areas can also be linked to distinct Software Engineering sub-zones.

i. Programming Testing
ii. Displaying and Specification
iii. Documentation
iv. d)Task Management e) Programming Quality
v. Programming Configuration Management
vi. Web Engineering are just a few of them.

7.5 Recommender Systems

Software engineering-specific recommendation systems are emerging to aid developers in a variety of tasks. Various recommendation systems assist people in making judgments and locating information in situations when they lack experience or are unable to analyse all available data. These systems integrate a variety of engineering and computer science techniques to make proactive suggestions that are tailored to the users' specific information needs and preferences. 1 Until now, the majority of recommendation systems were linked to the Internet. A large number of them, like Amazon.com's recommenders, are examples of mature technology provided as an element of commercial systems. The difficulties humans have while exploring enormous information spaces are similar to those faced by software developers trying to discover the one class they require among hundreds.

Recommendation Systems for Software Engineering (RSSEs) is gaining traction as a way to help developers with everything from code reuse to problem reporting. RSSE development is driven by the rising velocity of growth and heterogeneity of software systems and libraries, in addition to their magnitude. Similarly, when distributed programming becomes more common, information sharing among team members is becoming more difficult, motivating technology solutions. Large stores of currently accessible source code for mature software repository mining techniques, widespread adoption of common software development interfaces, such as Web interfaces like Bugzilla and tool-integration platforms like Eclipse, analysing recommendations, and, are all key factors in the development of practical RSSEs. RSSEs are ready to enter the toolkits of industrial software developers. Prototypes are evolving swiftly, tools are being published, and first-generation systems are being reimplemented in new settings. Recommendation systems assist consumers in overcoming information overload by exposing them to the most interesting objects and providing relevance, surprise, and innovation. RSSEs are similar to this description in that they try to assist developers in making decisions. RSSEs enable developers to discover the Application Programming Interface (API), human experts and proper code, in information spaces that include a system's code base, other documentation, libraries, version history, and, bug reports especially when it comes to information-seeking goals. RSSEs also tailor their output to the preferences of their users. Developers could communicate their interests either directly through a query or implicitly through behaviours that the RSSE considers when making suggestions. Such distinction illustrates a common problem for recommendation systems: determining context, which may include all relevant data about the user, her or his work conditions, and the project or task condition just at the time of the recommendations.

7.6 Internet of Things (IoT) Devices

In the past, software engineering techniques can be harnessed and adapted to the challenges of today's IoT. Nevertheless, new approaches to standard software engineering techniques are also needed—for example, rethinking configuration management in the context of the extremely dynamic, continuously reconfiguring systems that are characteristic of the IoT. A new generation of development environments is needed for software engineering for the IoT.

7.7 Blockchain Technology

Blockchain is used for data traversal in peer-to-peer networks and data storage in transparent ledgers because it is extremely secure. Growth in the number of blockchain-oriented apps has resulted from the increased availability of mobile applications with improved security and quality. The following are the primary characteristics of blockchain-oriented software applications (BOS) that assure data security:

a) Data Replication: The blockchain code is duplicated in each node. Data security is ensured since the data is replicated and stored across thousands of systems.
b) Transaction Recording: Transactions are recorded in a sequential log of interconnected blocks established by a consensus process in Blockchain-Oriented Software Systems (BOS).
c) Requirement Examines: BOS checks the transaction requirements before submitting them for validation.
d) Public-Key Cryptography: Public-key cryptography is a cryptographic program that utilizes pairs of private and public keys to secure transactions.

7.8 Digital Twin Technology

Processes and equipment are so complicated in today's world that the risk of failure or disruption through experimenting with new ways is either too great or too expensive. And this can be aggravating when new concepts have the potential to improve existing systems significantly. A digital twin is a virtual representation of a physical object. It uses analytics and IoT sensors to produce data-driven models of physical systems. Companies create digital twins by embedding sensors in their products and equipment to track and model system dynamics. Companies across various industries are using digital twin technology, along with the newest artificial intelligence and machine learning capabilities, to improve performance increase efficiency, cut operational costs, and revolutionise the way predictive maintenance is done. Digital twin technology is critical for product producers to have more efficient production lines and shorter time-to-market. Now, a few benefits of digital twin software are included here as:

a) Reduced costs. Before a functioning prototype is shown, a product usually goes through multiple revisions. It is quite expensive because the process necessitates a large amount of time and effort. Engineers can use digital twins to run tests and simulations inside a virtual environment, which decreases faults during actual production. In the digital realm, it is considerably faster and cheaper to fix flaws than it is in the physical world. Manufacturers can virtually eliminate all future output hazards and ensure that the physical thing will function exactly as intended.
b) Less time to market for new products. Due to numerous phases and periodic changes in production, getting to market faster than competitors is frequently a challenge. Digital twins' advanced image algorithms enable them to significantly cut time to market. The product life cycle takes place in a digital environment, which allows for quick and easy adjustments. A virtual prototype verifies how well the physical replica will operate in real life, reducing development time and increasing efficiency.
c) Predictive maintenance. To anticipate and address problems well in advance is the fundamental benefit of the digital twin technology. This characteristic is called as predictive maintenance. Virtual prototypes enforce consistent remote control of corresponding physical prototypes, collecting data from a variety of sources via sensors. The acquired data can be analysed to identify potential difficulties, such as when a spare part is nearly worn out and has to be replaced.

7.9 Quantum Computing

The fascination with quantum computing has grown tremendously in recent years. New reports regarding breakthroughs in this subject are released on a regular basis by research institutes, companies, and governments. Simultaneously, non-technical pieces discuss the probable repercussions of quantum computing, which range from cracking most current cryptographic techniques to a solution for complete general

AI. However, not all expectations are equally realistic. Apparently, we are several years away from having production-ready quantum computer hardware.

However, the broad concepts are already established, and abstractions help companies to use simulators to construct programmes that make use of quantum computing. Using traditional computers, conventional software development turns a high-level programming language (for example, Java) into processes on a huge number of (hardware) transistors. Quantum computing may very likely be able to tackle many problems that have previously been difficult to solve in practical terms. Because of its numerous possible uses, quantum computing is currently influencing most corporate sectors and research domains. Quantum algorithms must be explicitly programmed for these vastly distinct processors in order for such applications to become a reality. Despite the fact that several well-known quantum algorithms already exist, the demand for quantum software will skyrocket in the coming years. Quantum software must be generated in a more industrial and controlled manner in this setting, which means that issues like quality, delivery, project management, and quantum software evolution must be considered. We are confident that quantum computing will be the driving force behind a new software engineering golden era in the 2020s.

7.10 Quantum Machine Learning

Quantum computing's development and future abilities are expected to hasten improvements in existing computer paradigms, along with AI. Quantum machine learning is an emerging topic that will build quantum algorithms to do complex machine learning tasks. It is far more than just a conceptual combination of technologies. While Intel, IBM, Google, along with other large technology companies have lately made substantial advances in quantum computing, the technology still faces a number of challenges (not all of which are technological) before becoming a viable business resource. Machine learning and artificial intelligence (AI), as well as verifying and validating increasingly complex software systems, are among the computationally hard software engineering problems that the Department of Defense (DoD) faces. Finding the best answer to these problems, termed as combinatorial optimization problems, is non-deterministic polynomial hard and could take billions of years to complete using traditional computer paradigms. They are using quantum computing at the SEI's Emerging Technology Center (ETC) to solve these DoD mission-critical problems.

Our most recent project focuses on near-term quantum computing enabling software validation and verification within the Department of Defense. The integrated circuit computing paradigm has reached its fundamental limits since the start of the post-Dennard scaling era and the end of Moore's Law. Simultaneously, the Department of Defense (DoD) requires a new computing paradigm to aid in gaining a strategic advantage with machine learning and artificial intelligence (ML/AI)."The Pentagon is especially fascinated by the potency of quantum computing to grow secure communications and inertial navigation in GPS denied and contested environments," Michael Hayduk, chief of the communications and computing division at the Air Force Research Laboratory, told the Defense Innovation Board in July 2018. While we continue investigating these issues, we plan to expand our research into other areas, such as quantum machine learning, which uses quantum algorithms to conduct artificial intelligence and machine learning tasks. Researchers are also working on quantum interactive proof systems, forming interactive proof systems with QPUs, and validating and verifying quantum computation.

8. OPEN ISSUES AND CRITICAL CHALLENGES TOWARDS ARTIFICIAL INTELLIGENCE BASED SOFTWARE DEVELOPMENT

The majority of conventional, traditional software development environments adhere to the standard steps of analysis, design, planning, quality assurance, development, and deployment. The artificial intelligence development environment, on the other hand, operates differently. The creation of AI projects and the development is focused on identifying data sources, gathering data, cleaning it, and converting it into insights. A distinct attitude and skill set are required for such an approach. This unconventionality, which is woven throughout AI projects, brings with it a whole new myriad of issues and solutions for overcoming AI development obstacles.

The work exposures have rendered a few factors appear to us:

i. We cannot anticipate our AI software development project to produce the same results as a traditional product, because AI is more of a hit-or-miss game.
ii. When the entire team, not just the techies, is on board, we will be able to adopt AI strategies and programmes more effectively in our company.
iii. The constraints of AI initiatives, like those of non-AI app projects, differ from one idea to the next. However, some AI development issues and solutions are shared by all products.

Now some other challenges towards AI based software development could be:

a) Integration challenges: Integrating or adding, Artificial Intelligence into any current system is far more difficult than installing a browser plugin. There are a number of interfaces and aspects that must be set up in order to meet our company's requirements. The team of data scientists evaluates our particular data infrastructure demands, storage, data labelling and the processing of feeding data into the system, such that business would not have to deal with any startup AI app implementation issues. They additionally work on training the model and evaluating the AI's efficacy, creating a feedback mechanism for refining the models based on human activities.

b) Infrastructure capabilities: For businesses to deploy AI solutions, they must be able to handle data and compute it, as well as scale it, store it, extend it and secure it. When deploying an AI solution, a company's success begins with determining how to fit its infrastructure environment is and how effectively it supports AI applications and workloads. Unfortunately, the answer is also one of the most critical enterprise AI challenges. When it comes to the third lesson, there are challenges that are common to all products, regardless of which principle underpins them. We ran into similar issues no matter whatever application we were working on, so it's safe to presume they're common. To instil a proactive mindset in data engineers, entrepreneurs have compiled a list of the most typical concerns encountered when implementing AI development services, as well as their perspectives on each artificial intelligence opportunity and challenge.

c) Inefficient computing: Artificial intelligence necessitates the use of highly efficient and advanced machinery and processing. Cloud computing appears to be one option, but when we analyse current devices and software, it is clear that this is insufficient. One of the first hurdles that Artificial Intelligence technologies must overcome is this. Deep learning and Machine learning are two AI approaches that necessitate top-notch calculation speeds. For these, calculations must be performed at a microsecond or even nanosecond rate. In some circumstances, the calculation speed may need to be less than nanoseconds.

d) Lack of Support: The development of AI software is hampered by this lack of support. This is due to the fact that few people are aware of what artificial intelligence is, and even fewer are aware of how to utilise a machine capable of learning and thinking for itself. The rejection received from the public is what prevents it from progressing and reaching new levels of growth. As a result of the lack of demand from the public, there is no market for it, and as a result, firms and organisations do not invest in AI. This is how it deals with a lack of assistance.

e) Single-purpose specialization: Artificial intelligence is only been able to fulfil a few particular purposes thus far. It works by reading and retaining the inputs provided. However, it is restricted to improving only one task at a time. Artificial intelligence that can execute any task like humans still have not been achieved effectively. And enterprise mobility management necessitates this.

Although it might be created in the near future, it is currently unavailable on the market.

f) Biasion of Algorithms: AI programmes often operate based on the training they received from previous data. The issue develops when faulty data is introduced and the AI application begins to operate based on it. As a result, they must be trained upon unbiased data and create algorithms that are simple to understand.

g) Scarcity of Data: Even though organisations and businesses have enormous quantities of data, the data required for artificial intelligence is still insufficient. Furthermore, the most effective artificial intelligence would be one that undergoes supervised training, although this sort of training is based on labelled data, which is increasingly scarce in reality. As a result, artificial intelligence applications and machine learning systems that can accomplish more with fewer data are required to be developed and created. Also, perhaps, over time, the world would be able to generate sufficient data sets for Artificial Intelligence and machine learning systems to operate on, which is uncommon in today's environment.

Hence, this section has discussed several critical issues and open challenges towards AI based Software development. Now, next section will discuss several future research opportunities towards the same.

9. FUTURE RESEARCH OPPORTUNITIES

The emerging developments in artificial intelligence and software engineering are to model real-world objects such as expert knowledge, process models or business processes, and areas of research such as knowledge-based systems, agent-oriented software engineering, automated software engineering, and computational intelligence, are becoming increasingly important in both fields. Business intelligence is the practice of incorporating intelligence into software engineering processes, resulting in software development process automation. Software engineering data mining will aid in extracting important knowledge for identifying and selecting possible candidates for reusing, including using artificial intelligence techniques will aid in incorporating intelligence into the above-mentioned process. As a result, the whole domain of software engineering operations including software reuse will be automated. In businesses, Software Intelligence is referred to as Business Intelligence (BI) since BI aids SI in the software industry. The field of software intelligence can be researched for repurposing standard BI platforms for large-scale adoption of BI infrastructures.

i. Language APIs are used by 55.9% of AI and machine learning developers, preceded by speech APIs (51.1 per cent). For machine learning development and AI, developers have traditionally relied on a variety of APIs. The prominence of conversation and data discovery APIs is intriguing as demonstrates that voice-activated assistants are today integral to mainstream machine learning and AI software development.
ii. Today's Machine Learning and AI app development are being slowed by a lack of quality tools. The expense of resources, as well as a lack of requisite skills or training, are the most important challenges that AI and machine learning developers encounter in enhancing AI app development. Only 10% of respondents are dealing with the difficulties of working with and integrating legacy systems, indicating that AI and machine learning app development is taking place in relatively young business units and development centres.
iii. When it comes to designing AI applications, the complexity of managing operations is the greatest issue, according to 38% of AI developers. The development of programmes that are portable across deployment contexts is the second most critical problem. The third most difficult problem they confront in developing excellent AI products is deciding on the correct AI framework. Functions for different machine learning and deep learning procedures and Libraries of mathematical expressions make up AI and machine learning frameworks. They frequently offer a large number of APIs as well as other development tools to aid developers in incorporating existing code and leveraging business systems for the data required for training models and creating the app.
iv. For hosting their AI, machine learning, and deep learning projects, the majority of developers (54.9 percent) choose private cloud infrastructure. The public cloud is used by 46% of respondents, while organizations' on-premise infrastructure is used by slightly over 51% of respondents. Numerous cloud service vendors have constructed environments that include a variety of common AI tools, such as machine learning or deep learning frameworks, data science-specific IDEs, and machine learning notebooks.

10. SMART ERA WITH AI BASED SOFTWARE DEVELOPMENT: AN OPEN DISCUSSION

The development of AI-influenced systems incorporating AI capabilities driven by advancements in DL and ML has ushered in a new generation of AI in the software industry. For example, AI has taken a number of critical applications reaching a near-human degree of proficiency, such as autonomous vehicle navigation and speech and picture recognition. Software systems with AI components are known as AI-based systems. Such arrangements of systems learn via constantly understanding and analysing their surroundings and responding in order to achieve intelligent behaviour. The operation, development, and maintenance of AI-based systems differ from the maintenance and development of traditional software systems. System behaviour and rules are deduced from training data rather than being written down as code in AI-based systems. The advancement of AI-based systems necessitates a focus on changing and big datasets, an evolving and resilient infrastructure, as well as equitable and ethical engineering. We may wind up with unsatisfactory AI-based systems with technical debt if we don't acknowledge these differences. Under this perspective, it is necessary to investigate Software Engineering (SE) approaches for the maintenance, evolution and development of AI-based systems.

10.1 Software 2.0

The most significant change is the shift toward systems that have no ties to traditional coding at all. Neural networks and deep learning are examples of AI technologies that turn traditional software processes on its head. Developers, for example, begin with data instead of starting rather with rules — for an instance, enormous numbers of games. Google used AlphaGo for training the systems of tens of thousands of human games. The training data for the most recent version of AlphaGo Zero, comprised games that the system played in opposition to itself, beginning with randomized moves. This method has the potential to revolutionise software development as long as the training data is explicit and sufficient, and the criteria for success or failure are equally unambiguous.

Instead of attempting to decipher and code the game's rules, developers must now focus on success criteria and managing training data, leaving the actual programming for the system. Andrej Karpathy, Tesla's director of AI, says that's exactly the approach the company is adopting with its self-driving cars. Last year, he declared in a keynote at a technology conference, "This is a totally new method towards designing software." "Rather than explicitly writing code, we are now processing and accumulating data sets, which are effectively their code." For an instance, when driving through tunnels, Tesla's windshield wipers had problems determining when to switch on and off.

In the conventional days of software development, programmers would typically read and analyse the code and check for flawed logic. Instead of looking at the code, the engineers in Software 2.0 check the data. For example, there wasn't sufficient training set for vehicles travelling through tunnels. To deal with this shortcoming, Tesla collected new data, label it, added it to the training data set, processed it and executed the deep learning models again. However, as more businesses gravitate towards AI for applications with a lot of data and low-code infrastructures for the remainder, software development will undergo a significant transformation in the near future.

10.2 Smart Application Development Platforms

For as long as there has been development, developers have been connecting systems. Take, for example, Mendix, where the low-code approach is becoming even more adaptive, allowing organisations to avoid spending time on constructing commodity systems. For over a decade, the company has provided a building-block approach for designing applications. Developers assemble together functionality from the platform's options and, when they aren't adequate, link to external code to fill in the gaps. Today, the company has developed its deep learning network to evaluate these models, determine which ones have been the most successful, and detect patterns dependent on those models. Even the less creative aspects of the design process, which includes deciding on the best option for each stage, are prone to errors. As a result, smart specialists can help improve the traditional design process. For example, website construction platform Bookmark's Artificial Intelligence Design Assistant (AIDA) employs AI to evaluate users' wants and then generates a particular website for that sort of user. The system can choose from a myriad of combinations to design the website's look and feel focus, and other sections which could be customized. AIDA can generate a first version of the pages in less than 2 minutes, after which the design team can continue working on the website.

10.3 Developing Code

Whenever it comes to developing new code as from the ground up, today's technology leaves something to be desired. Microsoft's Visual Studio, the most popular IDE, now includes AI-assisted code completion in its current version. According to Mark Wilson-Thomas, Microsoft's senior product manager for Visual Studio IntelliCode, the feature is built on machine learning from hundreds of open-source GitHub sources. Amanda Silver, Microsoft's partner director of programme management for Visual Studio and Visual Studio Code, explains, "We extract the wisdom of the open-source community's code." "In a current study of IntelliCode customers, more than 70% said that the new AI-powered IntelliCode made them feel more productive than the original IntelliSense," she explains. The life of a developer entails more than just creating decoupled, clean code that produces user-friendly, interactive solutions. It also necessitates a significant amount of time, effort and resources spent, troubleshooting, resolving errors and reading documentation. Developers may immediately obtain feedback and ideas for improvement for the code they've written owing to smart coding assistants and machine learning, which saves them a lot more time. Python's Kite and Java's Codotaare two instances of such tools, but Microsoft's IntelliSense, which comes included with Visual Studio and is well integrated, is perhaps the most well-known at the time. Another way machine learning may help developers and businesses save time and money is through analysing the code more quickly and correctly and finding possible refactoring areas. However, identifying what typical patterns are good or harmful is where AI and ML may help on the development side. It may draw attention to the fact that this is an abnormality, allowing us to correct it later.

11. CONCLUSION

Artificial Intelligence, in combination with Machine Learning, Deep Learning, and Natural Language Processing (NLP), has the potential to completely revolutionize the process of software development. In layman's terms, AI-assisted technologies imitate human behaviour and do some routine tasks with extreme efficiency. Software developers and testers are becoming more successful and productive as AI and machine learning-powered technologies enable them to create high-quality software solutions. AI and machine learning not only aid companies in developing software, but also in understanding consumer behaviour, testing code, ensuring code security, and determining strategic choices. In light of this, IT executives must choose AI-powered tools and technologies to revolutionise their software development processes.

Conflict of Intertest: The authors have declared that they do not have any conflict of interest regarding publication of this work.

REFERENCES

Alhusain, S., Coupland, S., John, R., & Kavanagh, M. (2013). Towards machine learning based design pattern recognition. In *2013 13th UK Workshop on Computational Intelligence (UKCI)*. IEEE. 10.1109/UKCI.2013.6651312

Athavale, S., & Balaraman, V. (2013). Human behavioral modeling for enhanced software Project Management. In *7th International Conference on Software Engineering*, (pp. 15–7). IEEE.

Boehm, B., & Sullivan, K. (2000). *Software Economics: A Roadmap.* . doi:10.1145/336512.336584

Caldas, R. D., Rodrigues, A., Gil, E. B., Rodrigues, G. N., Vogel, T., & Pelliccione, P. (2020). A hybrid approach combining control theory and AI for engineering self-adaptive systems. In *Proceedings of the IEEE/ACM 15th International Symposium on Software Engineering for Adaptive and Self-Managing Systems*, (pp. 9-19). IEEE. 10.1145/3387939.3391595

Collobert, R., Weston, J., Bottou, L., Karlen, M., Kavukcuoglu, K., & Kuksa, P. (2011). Natural language processing (almost) from scratch. *Journal of Machine Learning Research*, *12*(Aug), 2493–2537.

Fenton, N., Hearty, P., Neil, M., & Radlinski, L. (2010). Software project and quality modelling using Bayesian networks. In *Artificial intelligence applications for improved software engineering development: New prospects* (pp. 1–25). IGI Global. doi:10.4018/978-1-60566-758-4.ch001

Filieri, A., Hoffmann, H., & Maggio, M. (2015). Automated multi-objective control for self-adaptive software design. *Proceedings of the 2015 10th Joint Meeting on Foundations of Software Engineering*. 10.1145/2786805.2786833

Gao, H., Peng, Y., Jia, K., Wen, Z., & Li, H. (2015). Cyber-Physical Systems Testbed Based on Cloud Computing and Software Defined Network. In *2015 International Conference on Intelligent Information Hiding and Multimedia Signal Processing (IIH-MSP)*. IEEE. 10.1109/IIH-MSP.2015.50

Ghezzi, C., Jazayeri, M., & Mandrioli D. (2002). *Fundamentals of software engineering*. Prentice Hall PTR.

Horn, G., Skrzypek, P., Materka, K., & Przeździęk, T. (2019). Cost benefits of multi-cloud deployment of dynamic computational intelligence applications. In *Workshops of the International Conference on Advanced Information Networking and Applications*, (pp. 1041-1054). Springer. 10.1007/978-3-030-15035-8_102

Jarrahi, M. H. (2018). Artificial intelligence and the future of work: human-AI symbiosis in organizational decision making. *Business Horizons*, *61*(4), 577–586. doi:10.1016/j.bushor.2018.03.007

Kakatkar, C., Bilgram, V., & Füller, J. (2020). Innovation analytics: Leveraging artificial intelligence in the innovation process. *Business Horizons*, *63*(2), 171–181. doi:10.1016/j.bushor.2019.10.006

Laplante, P. A. (2000). *Dictionary of computer science, Engineering and Technology*. CRC Press.

Leau, Y. B., Loo, W. K., Tham, W. Y., & Tan, S. F. (2012). Software development life cycle AGILE vs traditional approaches. *Int Conf Inf Netw Technol.*, *37*(1), 162–167.

Mayhew, D. J. (1999). The usability engineering lifecycle. In CHI'99 Extended Abstracts on Human Factors in Computing Systems (pp. 147–8). ACM. doi:10.1145/632716.632805

Meinke, K., & Bennaceur, A. (2018). Machine learning for software engineering: models, methods, and applications. In *2018 IEEE/ACM 40th International Conference on Software Engineering: Companion (ICSE-Companion)*. IEEE. 10.1145/3183440.3183461

Muenchaisri P. (2019). *Literature reviews on applying artificial intelligence/machine learning to software engineering research problems: preliminary*. Academic Press.

Oliveto, R., Antoniol, G., Marcus, A., & Hayes, J. (2007). Software Artefact Traceability: The Never-Ending Challenge. In *2007 IEEE International Conference on Software Maintenance* (pp. 485–488). IEEE. 10.1109/ICSM.2007.4362664

Padmanaban, P.H., & Sharma, Y.K. (2019). Implication of artificial intelligence in software development life cycle: a state-of-the-art review. *IJRRA*.

Poole, D. L., & Mackworth, A. K. (2010). *Artificial intelligence: foundations of computational agents*. Cambridge University Press. doi:10.1017/CBO9780511794797

Ramadan, R., & Widyani, Y. (2013). Game development life cycle guidelines. In *2013 International Conference on Advanced Computer Science and Information Systems (ICACSIS)*. IEEE.

Ruparelia, N. B. (2010). Software development lifecycle models. *Software Engineering Notes*, *35*(3), 8–13. doi:10.1145/1764810.1764814

Savchenko, D., Kasurinen, J., & Taipale, O. (2019). Smart tools in software engineering: a systematic mapping study. In *2019 42nd International Convention on Information and Communication Technology, Electronics and Microelectronics (MIPRO)*. IEEE. 10.23919/MIPRO.2019.8756975

Seffah, A., Gulliksen, J., & Desmarais, M. C. (2005). Human-centered software engineering-integrating usability in the software development lifecycle (Vol. 8). Luxemburg: Springer Science & Business Media.

Sorte, B.W., Joshi, P.P., & Jagtap, V. (2015). Use of artificial intelligence in software development life cycle—a state of the art review. *International Journal of Advanced Engineering and Global Technology*, 398–403.

Stylianou, C., & Andreou, A. S. (2016). Investigating the impact of developer productivity, task interdependence type and communication overhead in a multi-objective optimization approach for software project planning. *Advances in Engineering Software*, *98*, 79–96. doi:10.1016/j.advengsoft.2016.04.001

Van Hoorn, A., Frey, S., Goerigk, W., Hasselbring, W., Knoche, H., Köster, S., & Wittmüss, N. (2011). *Lübeck: Dynamic analysis for model-driven software modernization*. DynaMod project.

Velmourougan, S., Dhavachelvan, P., Baskaran, R., & Ravikumar, B. (2014). Software development life cycle model to build software applications with usability. In *2014 International Conference on Advances in Computing, Communications and Informatics (ICACCI)*. IEEE. 10.1109/ICACCI.2014.6968610

Zhang, H., Feng, M., Long, K., Karagiannidis, G. K., & Nallanathan, A. (2019). Artificial intelligence-based resource allocation in ultradense networks: Applying event-triggered Q-learning algorithms. *IEEE Vehicular Technology Magazine*, *14*(4), 56–63. doi:10.1109/MVT.2019.2938328

Zhang, Y., Robinson, D. K., Porter, A. L., Zhu, D., Zhang, G., & Lu, J. (2016). Technology roadmapping for competitive technical intelligence. *Technological Forecasting and Social Change*, *110*, 175–186. doi:10.1016/j.techfore.2015.11.029

Chapter 4
The Influence of the Evolution of Technology 4.0 in Continuing Education

Ramiro Reyna
Instituto Internacional de Estudios Superiores, Mexico

ABSTRACT

The technological advances that the world has experienced throughout all the years have caused important changes in the structures of society. Education is one of the fields in which the industrial revolution, driven by technological development, caused a significant impact, since technology allowed students to freely access more information and therefore more knowledge. Education 2.0 and 3.0 brought not only access to information, but also greater connectivity and interaction between people, particularly between teachers and their students. The dynamism of technology also brought opportunities for innovation with Education 4.0, changing the existing paradigm from face-to-face to online classes, the emergence of digital, your relationship with those around you, and your abilities to meet basic demands and roles in society. These changes meant the application of new educational methodologies, aimed at significant learning that prepares the individual with the necessary skills to face the modern world of work.

DOI: 10.4018/978-1-6684-8088-5.ch004

INTRODUCTION

There is no revolution without a traditional break into the future. Its consequences destroy traditions, behaviors, relationships, ways of thinking and, of course, of educating. Throughout time, societies have undergone great transformations due to the advancement of technology, which brought with it important changes in their structures and altered the existing statism. Since the industrial revolution, education began to contribute to the economic development of countries, focusing on pedagogy for the working class and working class who maintain the economy of their countries or entities.

Technological development driven by industrial revolutions has a direct impact on education. Thus, Education 2.0 transformed memory education when the student began to have access to information and greater knowledge, all thanks to the use and application of emerging technologies. With the globalization of the Internet, the generation of greater forms of research, and the variety of knowledge and sources, the student was immersed in Education 3.0. Virtual platforms appeared, connectivity and interaction between students and teachers increased, and a more personalized and reflective learning model was applied. On the other hand, the key element of Education 4.0 is innovation, and it has radically changed the way in which individuals live, work and relate to others.

Today's world is experiencing a moment of important transformations. The learning activities carried out by the students, as well as the application of methodologies and techniques by their tutors, constitute the basis of a dynamic and changing digital or face-to-face educational model. This innovation process Structural has bet over time for the development and transformation of science and technology, until it evolved into the current Education 4.0. An important aspect is the development of students' skills during the learning process, which allows them to be able to enter the labor market and meet the labor competency requirements of companies. In this way, society will be able to count on a labor force that, in addition to fulfilling its basic roles, is proactive, resourceful, creative, and innovative.

Technology began to have a place in educational processes in a basic way at the beginning of the new millennium, and it became known as Education 2.0. At this stage, the teacher not only dedicated himself to teaching, but also acted as a moderator to promote interaction in the classroom, while discovering the talents of his students through collaborative work.

On the other hand, Education 3.0 has made it possible to incorporate Information and Communication Technologies (ICT) into educational processes, thus integrating new tools that have improved the way of educating in recent years. Education established a closer connection with different sources of information. Learning was self-directed and the search for information was encouraged in a 100% digital way. This new concept stimulated the students' skills, guiding them to create new content based on research, which significantly helped both students and teachers to change their mindset where the interaction became more dynamic, participatory, and creative.

Among the benefits of the implementation of ICT in education are:

- Access to information and education from any device regardless of location.
- Develop skills that allow us to function in a digital environment.
- Immediacy, communicate in real time.
- Low cost of education, both for students and for institutions.
- Education adapted to the needs of the students.
- ICTs have made it possible to communicate in real time
- Improvement of the knowledge acquired in the groups of students.
- Development of new skills outside the official curriculum.

PERSONALIZED LEARNING

The arrival of Education 4.0 represents an ideal approach, aligning the Fourth Industrial Revolution with education; This is because the industrial revolution bases its development on robotics, smart technology, artificial intelligence, augmented reality, gamification, teacher evaluation and student self-assessment, big data, use of digital platforms; facilitating our daily life (Jabbar et al. 2023)

Universities are challenged to produce successful professionals using these new ways of teaching, where students and professors live together in an environment that is up to Industry 4.0 and where physical cybernetic systems are applied in all industries. Part of the curriculum should always include these technologies. In this way, the learning approach will always be based on new ways of studying and their application in new work environments.

Through Education 4.0, the desirable skills of professionals can be improved, turning them into innovative and creative employees; with the ability to adapt to the use of new technologies and corporate strategy of companies.

Some of the characteristics of Education 4.0 are the following:

- Adaptive learning to the profile of each student.
- Self-managed learning.
- Flexible learning, linked to the productive sector.
- Constant feedback, based on data analysis.
- Communication as the main link of learning.

- Cooperation between students and teachers.
- Learning through educational competencies model.
- Evaluation as a constant process to improve and progress.

RESOLUTION OF REAL PROBLEMS

Use ICT as tools for access, organization, creation and dissemination of content. The evolution of education can be seen through a time line. The contributions to the social structure and superstructure of the state of the First Industrial Revolution that took place in England at the end of the 18th century created a new model of life, causing transformations in the pillars economic, political, social, cultural and environmental aspects of society. countries. An educational model emerged that primarily considered the needs of society and whose premise was economic development. The schools were the engines that supplied the professionals of an industrial society, whose base was mass production and consumption, to facilitate the development of their skills and abilities. This gave rise to the ability of the human being to adapt to the needs and advances of science.

Current advances in science and technology are part of the Fourth Industrial Revolution, which developed throughout the year 2000 and is also known as Education 4.0. Electronic commerce and cellular communication led society to enter a world of technological challenges at an accelerated pace, with astonishing levels of speed of information and sending and receiving data. The development of new information and communication technologies arose, such as the so-called disruptive technologies, among which are applications (Apps), intelligent systems and digital production processes, facilitating the way in which the individual learns and teaches to the new generations.

Similarly, production processes are standardized and systematized thanks to technologies that mark innovation, such as robotics, virtual reality, artificial intelligence, IoT or Internet of Things, gamification, or Big data, among other technologies., which are impressively marking technological innovation in various economic fields. sectors worldwide, such as the industrial sector, commerce or communications, as well as in the way of doing business, in politics or in marketing.

In each industrial revolution, knowledge transfer has been required, and currently in Industry 4.0 it is necessary to carry out a digital transformation process. The implementation of new technologies requires the redesign of the business architecture, to work by processes, solutions and software applications, mobile, remote, with the use of data from appropriate sources, integration of information and communication technologies.

Accordingly, the Traditional Educational System has been characterized by transmitting content to students, with a rote learning methodology without further development of creativity, development of skills and abilities of students, who are evaluated through tests that are mainly written. . and oral, to have the supposed evidence of learning. It should be noted that this education has the teacher as the axis of the teaching-learning process, without allowing greater feedback on the knowledge and skills of their students, who are passively instructed.

In practice, it has been scientifically proven that long-term memory is not achieved by memorizing efforts in the moment. What is not applied by students, according to technological advances and the development of their life daily, it is the memory that is lost, limiting meaningful and lasting learning.

Traditional education evolved towards an education with feedback processes so that the student had access to various sources of knowledge consultation and could contrast the information provided by their teacher. The student went from being just a receiver of information to a generator of ideas from the compendium of acquired knowledge, capable of sharing them through virtual means and also of globalizing knowledge. It is not enough for students to memorize the contents, since in the globalized world in which we live, anyone has easy and quick access to any content to investigate and learn. It is necessary to change the traditional rote education of contents, for an education that is constantly fed by knowledge and that, in the virtual, online modality, is dynamic, integrating the universality of knowledge. Education 4.

Face-to-face, blended and virtual education, with the use of educational applications and platforms, has allowed the generation of learning environments that make it easier for the tutor to implement efficient alternatives and promote new abilities and skills in students, inside and outside the classroom. The modern teacher influences the student so that they can be inserted in the 21st century education, education 4.0 and considering the rhythms and learning styles of the students, generate changes to be implemented in both public and private educational institutions, with redesign of strategies of teaching-learning within the student-teacher relationship.

The impact of the industrial revolution on education, particularly in Education 4.0, requires digital transformations along with the technological era, to meet the needs of the information society at the pace of scientific and technological transformations and innovations, framed in a new educational model. To achieve this, it is necessary to change the behavior and social customs of each culture. The co-responsibility of the teacher in the construction of this new model requires efforts to adapt to this dynamic, didactics, their own training and professionalization, the continuous acquisition of knowledge and all this with the use of technological tools inside and outside the classroom as a transverse axis.

Martinic and Villalta (2015) affirm that Education 4.0 is based on ten technological pillars, which are explained below:

Integration systems, which integrate operational technologies with information and communication technologies, impacting both the internal and external management of a company.

Robots, or intelligent systems that automate tasks previously performed by humans. This pillar seeks to strengthen and increase collaborative robotics.

Internet of Things (IoT), the same one that makes multidirectional communication between people, machines and products possible, through sensors that are combined with Cloud computing and Big data analysis, as stated by Ortiz, F (2023).

Additive manufacturing, through which it is possible to manufacture parts by superimposing layers of materials from a previous design. It does not use molds. Instead, it takes a virtual model.

Big data, characterized by the volume, speed, and variety of both structured and unstructured information, which can come from various sources and be reported by various means. The biggest advantage it presents is the generation of reports based on the data analysis process.

Cloud. This pillar offers online storage, access and use of IT services. In this way, companies access existing technologies with little administrative effort and in a flexible and agile way.

Simulation of virtual environments. This pillar has to do with a joint way in which machines, people and processes work directly. Prevent breakdowns, save money and evaluate the final results in a controlled environment.

Artificial intelligence. It refers to a set of algorithms fed with information and previous experiences, improving them to provide machines with cognitive abilities similar to those of the human brain.

Cyber security. Technological advances require the use of cybersecurity mechanisms in industrial environments, detecting, anticipating and neutralizing threats.

Augmented reality. It complements a real environment with a digital one, facilitating the design of products and organization processes.

The pillars of the Fourth Industrial Revolution are focused on the digitization of the manufacturing sector, driven by the large volumes of data that are handled, connectivity, and powerful computer systems. However, these disruptive technologies exert a strong influence on education and have changed the way teaching and learning is carried out. It is important to consider that these pillars are complemented by the offer of other existing technologies such as drones, controllers, platforms, and location and self-identification systems, which play as important a role as the aforementioned pillars in the expansion of current technology. matrix. The rapid spread of the COVID-19 pathology is closely linked to the vast network of global movement of people. Contagions were inevitable. The difference with the pathological facts of the past is that in the current century information circulates instantly, and in the case of the SAR virus, it proliferates at the same speed as the number of infected people increases. In the educational field, the massive closure of educational institutions worldwide to mitigate the spread of the virus raised an unprecedented risk to the education and well-being of the people. The closure of schools, colleges and universities also affected the nutrition of the student population, especially that belonging to vulnerable segments. The massive closure of educational institutions around the world to mitigate the spread of the virus posed an unprecedented risk to people's education and well-being. The closure of schools, colleges and universities also affected the nutrition of the student population, especially that belonging to vulnerable segments. The massive closure of educational institutions around the world to mitigate the spread of the virus posed an unprecedented risk to people's education and well-being. The closure of schools, colleges and universities also affected the nutrition of the student population, especially that belonging to vulnerable segments.

Despite the emergency, societies opted for technology to continue with the educational process, causing a change in people's way of life, but also revealing deficiencies in terms of infrastructure and teaching skills.

The transfer of technology and knowledge of Education 4.0 before the COVID-19 pandemic grew in a similar proportion to the levels of investment that governments allocated to their Gross Domestic Product (GDP) and to the variables of innovation, development and research in science and technology. In other words, if a country's investment levels were low, the development and implementation of Education 4.0 would not globalize beyond small business and academic spaces. However, in the transition from the pandemic, drastic changes occurred in face-to-face education, e-commerce and home sales.

Most of the countries decided to establish ways to continue the educational process in various distance modalities. Some chose to implement online forms of learning, while others decided to establish distance learning strategies in offline mode.

This caused a significant increase in the need for access to ICT, as well as the educational model inserted in Education 4.0, which includes smart technology, artificial intelligence and robotics to adapt to the new way of teaching and learning in the new business era. The students went from the face-to-face methodology to the virtual education methodology, finding themselves with an educational system in which the teachers were not prepared to assume another educational model and leave the traditional method.

Most of the countries were able to count on the adequate resources and digital platforms to establish a remote connection. In this case, many found themselves in the need to increase speed to unprecedented levels and implement open radio or television programs.

Societies around the world benefit from technology on a daily basis, since it provides a series of advantages, especially in the educational field where it has revolutionized the way of teaching and learning.

A model of pedagogical training that remains valid despite the arrival of Education 4.0 are the Massive Open Online Courses (MOOC), the same ones that appeared with Education 3.0. At the time, this model revolutionized the way in which teachers acquired knowledge and skills to improve their practices. According to Núñez-Peña et al., (2015); (Jiménez-Valverde & Núñez-Cruz, 2009) The first MOOC was implemented by the University of Manitoba, Canada, in 2009, with around 2,000 students enrolled. This platform had the feature of combining a Webinar or blog posts from professional experts with social media messages from Twitter from the course participants. These courses were open to anyone interested and did not include formal assessments. The author explains that, in 2012, two Stanford University professors offered an action-based MOOC course capturing lectures on AI, which attracted around 100,000 participants. William assures that from that moment on, MOOCs spread rapidly throughout the planet.

The e-learning education platforms have evolved so much that the environment that creates a virtual classroom and the virtual educational campus, provide facilities for the learning of the student of the new era and insert it to the rhythm of the world market. We are already experiencing the projection of job offers and the acceleration of the application of science and technology today, even as an effect of the confinement caused by the pandemic caused by Covid-19. There has been such a serious school dropout, caused by causes such as the lack of access to computer equipment, whether portable or desktop, the lack of Internet access, by senior teachers who refuse to join the school virtual education and therefore cannot apply pedagogies that allow students to enjoy virtual classes.

According to Rodriguez et al., (2014); Mendoza-Rojas, (2015) it is expected that there will be an increase in school dropout in the world of approximately 24 million children. According to the entity, only in Ecuador there are more than 90,000 students who have left the educational system, and approximately 15% of the students who have not left the system indicate that they have not had regular contact with teachers in at least two weeks. UNICEF stresses that it is important to consider that the loss of contact between teacher and student increases the possibility of the student dropping out, and that the experience of such a high number of dropouts only increases the 268,000 children who had already dropped out of school . educational system and the 187.3 thousand who were already behind in school.

In addition, another international entity, the World Bank (WB), reports an estimate of USD 10 trillion in lost income worldwide, due to lower educational levels, with the consequent risk of children dropping out of the educational system. According to this entity, there is a global worsening of the children's abilities in reading, writing and mathematics. Additionally, the WB mentions that, in Ecuador, 6 out of 10 students indicate that they are learning much less since the COVID-19 pandemic began. Therefore, it is concluded that, despite the increase in connectivity, with 74.8% of students with access to the public Internet, only 1 in 8 students have computers or some equipment that facilitates Internet access for use. staff. This reality, as the WB states, has a significant impact on students' online learning.

The year 2020 brought with it important changes, especially in the traditional forms of the teaching-learning process, leading the world population to include virtual education to remedy the limitations caused by the COVID-19 pandemic. This new way of working left important lessons learned, and the premise of continuing with the innovations that Education 4.0 has brought at a dizzying pace.

MATERIALS AND METHODS

To develop the present study, in the first instance, the bibliographic material was used (which serves to explicitly cite all the bibliographic references required in the elaboration of the theoretical framework, citing the authorship of each collaborator of the raised topic). The historical method was used as a second methodology in order to establish the bases to analyze the descriptive research methodology and thus be able to answer questions about the reality of the study carried out. This allowed establishing a before as a starting point and an after education in the present study. With this, changes in the teaching and learning methodology were evidenced throughout the 20th century to the present.

RESULTS

Pedagogical renewal and innovation, generally postponed, but implicit in Education 4.0, now have a great opportunity to become a reality and allow both quality and equity in education. Therefore, it is recommended that in the immediate future educational systems be implemented that consider technology as their main resource, that are resilient, that can respond efficiently to the emerging situations that this changing world presents, and that can face the challenge to reduce the inequalities that worsened like never before with the pandemic, through Education 4.0. The technological advances that the world is facing today, such as robotics, artificial intelligence, Big data or the Internet of Things (IoT), which have greatly impacted the world,

DISCUSSION

The drastic change in daily life caused by the COVID-19 pandemic, added to the general economic crisis, has meant a significant change in the way, the place and the moment in which the learning of the students takes place. The human being has had to quickly adapt to different ways of teaching and learning, and teachers have been forced to develop new abilities and skills, especially in the area of technology-oriented education, in order to maintain the quality of education and the profitability of the educational turn of private and government institutions.

Education 4.0 has had a great impact on education under the pandemic scenario that the world is experiencing, since it has allowed teachers to implement innovative strategies, and seek adequate resources to maintain student participation and motivation, thus highlighting the importance of this fourth industrial revolution.

The evolution of e-learning education platforms guarantees a suitable environment for the virtual classroom, and even for a virtual university campus, since it offers all the facilities that student learning needs in modern times, and inserts it at the pace of the world market.

REFERENCES

Jabbar, M. A. Tiwari, S., Ortiz-Rodriguez, F. (2023). Smart Urban Computing Applications. River Publishers.

Jiménez-Valverde, G., & Núñez-Cruz, E. (2009). Online cooperation in virtual environments in the teaching of chemistry. *Chemical Educator (Boise, Idaho)*, *20*(3), 314–319. doi:10.1016/S0187-893X(18)30031-4

Lopez, S. I. M., & Valenzuela, B. G. E. (2015). CHILDREN and adolescents with special educational needs. *Clínica Las Condes Medical Journal*, *26*(1), 42–51. doi:10.1016/j.rmclc.2015.02.004

Martinic, S., & Villalta, M. (2015). Classroom time management and school performance in full-time schools in Chile. *Educational Profiles*, *37*(147), 28–49. doi:10.1016/j.pe.2013.03.001

Marúm-Espinosa, E., & Reynoso-Cantú, E.-L. (2014). The importance of non-formal education for sustainable human development in Mexico. *Ibero-American Journal of Higher Education*, *5*(12), 137–155. doi:10.1016/S2007-2872(14)71947-X

Mendoza-Rojas, J. (2015). Expansion of the offer of higher education in Mexico and creation of public institutions 2001-2012. *Ibero-American Journal of Higher Education*, *6*(16), 3–32. doi:10.1016/j.rides.2015.04.001

Núñez-Peña, M., Bono, R., & Suárez-Pellicioni, M. (2015). Formative Assessment in Higher Education: Impact on Students with Math Anxiety. *Procedia - Behavioral and Social Sciences, 196* (July 2014), 135-141. doi:10.1016/j.sbspro.2015.07.023

Orozco, J. E. F., Becerra, J. I. V., & Arellano, C. I. M. (2015). Current Perspectives on World University Rankings. *The Journal of Higher Education*, *44*(175), 41–67. doi:10.1016/j.resu.2015.09.001

Ortiz-Rodriguez, F., Tiwari, S., Sicilia, M. A., & Nikiforova, A. (2023). *Electronic Governance with Emerging Technologies*. First International Conference, EGETC 2022, Tampico, Mexico. Revised Selected Papers.

Rodríguez, G., Ariza, M., & Ramos, J. L. (2014). Institutional quality and academic performance The case of the Colombian Caribbean universities. *Educational Profiles*, *36*(143), 10–29. doi:10.1016/S0185-2698(14)70607-5

Rojas, MS (2014). Violence against women in the news in Mexico: the uncritical and uneven treatment of a serious social problem1. *Sociological Act, 65*(2012), 37-64. . doi:10.1016/S0186-6028(14)70236-1

Tobón, O. E. A., Zapata, S. J. C., Lopera, I. C. P., & Duque, J. W. S. (2014). Academic training, values, empathy and socially responsible behaviors in university students. *The Journal of Higher Education*, *43*(169), 89–105. doi:10.1016/j.resu.2015.01.003

Chapter 5
Using Technologies to Track Products and Predict Failures

Mario Pineda
INDEX, Mexico

ABSTRACT

This chapter reviews and discusses three pervasive topics in current manufacturing processes around the world: 1) internet of things, considered a key factor in the Industry 4.0 revolution, and how its applications can be used to enhance supply chain and manufacturing operations; 2) serialization and tracking of manufactured products, which discusses the value and usefulness of product traceability for warranty purposes and to reduce counterfeiting; 3) ERP systems in manufacturing processes, which, when used appropriately, provide real-time data and analysis capabilities for instant and long-term decision-making, which in turn can provide a competitive advantage for companies. The author's goal is that this chapter may serve informational and referential purposes, and as a practical guide to either launch or implement real-world careers or processes in the areas discussed.

INTRODUCTION

This chapter reviews and discusses three pervasive topics in current manufacturing processes around the world: 1) Internet of Things, considered a key factor in the Industry 4.0 revolution, and how its applications can be used to enhance supply chain and manufacturing operations; 2) Serialization and Tracking of Manufactured Products, which discusses the value and usefulness of product traceability for warranty purposes and to reduce counterfeiting; 3) ERP systems in manufacturing processes, which, when used appropriately, provide real-time data and analysis capabilities

DOI: 10.4018/978-1-6684-8088-5.ch005

for instant and long-term decision making, which in turn can provide a competitive advantage for companies. My goal is that this chapter may serve informational and referential purposes, and as a practical guide to either launch or implement real-world careers or processes in the areas discussed.

IoT

The Internet of Things (IoT) (Ortiz-Rodriguez et al, 2023) is the collective network of computing mechanical devices that transfer data over the Internet. IoT encompasses home appliances, point-of-sale devices, mechanical devices, sensors, security devices, smart wearables, medical devices, personal transportation vehicles, etc. The IoT definition usually excludes traditional computational devices such as smart phones, computers, and servers. IIoT refers to the Industrial or Manufacturing version of IoT.

Current State of IoT:

As of 2022, (Fernandez et al, 2022) the global market is estimated at US$326 billion, and expected to hit US$1.3 trillion in 2032. The ever-increasing demand for sensors and machine upgrades has not been met by the supply. Even before Covid-19 hit, the semiconductor supply chain was already experiencing component shortages. During and after the pandemic, the issues exacerbated even further. Many companies, and industries even, have scaled down their throughput and their offerings to their customers due to these supply-chain issues.

One of the most significant limitations to the growth of IIoT market share is the ease with which alternative substitutes are available. The pandemic resulted in two major demand drivers that are further undercutting the demand from traditional industrial IoT: one is the increased demand for personal devices (computers, phones, etc.) from people who went to work from home; the other is related to remote- patient monitoring, where people would avoid clinics and would need devices to "visit" their health providers.

How Is IoT Relevant in The Manufacturing Industry?

Industrial IoT (IIoT) helps companies to automate and optimize their manufacturing and operational processes. Any device or machine in an industrial or manufacturing facility can either be upgraded or replaced with "smart" technology to enable it to become an IoT device. The intention is to connect that machine or device to the IoT network with the purpose of monitoring it and gathering its performance and test results data.

Once connected to the IoT network, the device or machine can help in reducing errors and the corresponding downtime and can help improve product quality. Cost reduction is also a mid-to-long term proposal through the prediction and prevention of corrective maintenance.

The topology of an IT-centric network has always been direct and straightforward, with the aim to be managed centrally. IIoT has resulted in a more de-centralized IoT because manufacturing operations require more stringent, faster and more response times, and larger sets of data storage and exchange. Machines, devices, sensors, etc. may create an immense amount of data (usually unstructured, and not transactional) that poses a challenged when attempted to be transmitted to a remote location, even if that remote location is the company's, IT headquarters or data center. An ideal IIoT network should be private, running on fast Ethernet or 5G broadband that would allow for greater coverage, control, flexibility, response, and speed.

Information Security (InfoSec) in an IIoT network sometimes differs from a traditional IoT centric network in that it may not be managed by IT personnel. In a manufacturing environment, login credentials, access to specific network resources, or file sharing is usually not managed with a mindset of InfoSec. It's not unusual to see Engineers or Technicians share passwords with one another across the cells in a production line. It's also not unusual to see Engineers develop software solutions tailored specifically to a machine or device, without much concern on IT security practices or software development standards.

Best Practices to Implement an IIoT Network

The increasing need for industrial data (Jabbar at al 2023) gathering and subsequent analysis has pushed operational technologies to attempt to match Information Technology offerings through IoT. Here are listed several best practices to serve as a guide for the design, implementation, maintenance, and application of IoT networks.

Best practice 1: gradual transition to IIoT. The road to IIoT adoption is full of unknowns, roadblocks (in the form of communication protocols, machine updates and sensor installations), user resistance, and the challenge of how to transmit, how to store, and how to analyze resulting data sets. Convert, or deploy new, one IIoT machine (or one set of identical machines) at a time. Then, monitor behavior, monitor results, analyze gathered data points, evaluate the complete installation, including challenges and costs, and repeat. If the upgrade of existing equipment or machines is too insurmountable challenge, consider setting up a parallel IIoT environment that would function for a pre-determined period of time in addition to the current setup. The old adage of "to do it well, do it slowly" aptly reflects this recommendation.

Best practice 2: select and stick to a single Messaging Protocol. Data gathering and transmission with the main goal to make decisions, is what companies expect to obtain from an IIoT network. IIoT protocols fall into two categories: 1) Publish-and-Subscribe (pub-sub) protocols that connect and publish data to a topic on an intermediary broker, such as MQTT, AMQP, DDS, and XMPP; 2) client-server or poll-response protocols, such as Allen-Bradley, Modbus, and Omron, where clients connect continuously to the server to verify if data has undergone changes.

Best practice 3: Implement Redundant Edge Devices. Industrial and manufacturing environments rely heavily on systems uptime. Machine connectivity can be either hardline or remote. Edge computing refers to groups of devices including multiplexers, routing switches, LAN / WAN devices, and routers. An edge gateway is basically a combination of a router, terminal server, network arbitrator, and network box. The term "edge" means that they reside on the periphery of SCADA systems and require failsafes in the form of redundant devices. A common best practice is to place a redundant edge device at every point where data is generated and/or collected.

Predictive Maintenance (PdM)

Equipment maintenance is a critical activity in any industry, and particularly in Manufacturing. Machine failures can lead to adverse results in meeting production goals such as productivity, delivery, cost control, and quality control. A consistent and comprehensive plan of continuous monitoring of equipment and machines to schedule preventive maintenance is a key component in a company's quest for continuous improvement.

IoT devices, when appropriately programmed, can self-detect a potential failure and report maintenance requirements. Through wireless sensors installed in the machine or device, a technician can be alerted to an upcoming preventive maintenance routing or even a potential failure that will necessitate corrective maintenance. By building on data provided by IoT-enabled sensors, PdM uses analytics and incipient technologies (such as Artificial Intelligence) (fig. 1) to predict potential failures before they occur and can also anticipate, and report needs for maintenance before they become necessary.

As real-world examples, Infineon provides sensors that can be deployed to monitor machine-specific components to entire smart ecosystems. Typical cases include HVAC equipment to monitor vibration, motor airflow, and current. When combined with microcontrollers and security solutions, the resulting collected sets of data can be processed and analyzed to enable advanced IoT features such as PdM.

Figure 1. The value chain for smart building technology

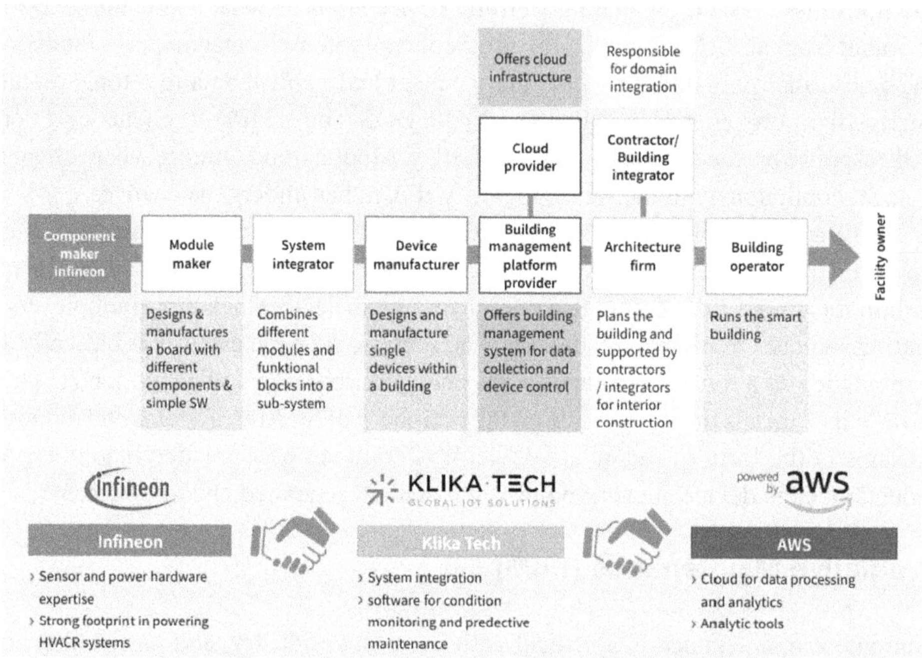

Remote Monitoring and Control

Not all machines are easily accessible for maintenance, and this is where remote monitoring and potential remote maintenance can be performed through the use of IoT. For example, IoT sensors can detect leakage in water pumps by alerting and activating shutdown procedures in real time. In regards to energy optimization, and as the world becomes more conscious of its energy consumption, the power needs of IoT connected machines can be adjusted automatically or remotely, during peak production or peak energy-usage periods.

Asset Tracking

With IoT, long gone are the days where a worker needed to be physically in front of a machine or device to confirm or record its existence, availability, and correct functioning. Modern supply chains rely heavily on asset-tracking systems to maintain correct inventory records, to optimize logistics and to ensure the safety of the assets.

In transportation, for example, a company's assets are constantly on the move in multiple geographical areas, be it on land, sea, or the air. IoT allows the constant monitoring and tracking of every vehicle and can assist in reporting potential breakdowns.

Smart Metering

Also long gone are the days where a technician would have to drive or work to every consumer or company's physical location to read its individual energy, water, or gas consumption. With smart metering technology, the Utilities providing those services are now able to remotely read consumption, reset the meter, adjust rates, connect/disconnect service, detect leakages, or detect wide-area failures.

Careers in IoT

The continuous, sustained growth in the IIoT arena provides exciting opportunities for individuals to explore, specialize, and thrive. We will explore a few careers to explore for those interested, but first, let's discuss how to begin a career in IIoT.

First, the candidate should have an understanding of, or at least a liking to, big data, data analysis, machine learning (ML) and artificial intelligence (AI). Both AI and ML assist devices and machines to make decisions in response to the data provided, and to identify potential failures.

Second, a hand-on experience with IoT and embedded devices is priceless because such experience provides the technical knowledge to interact with IIoT systems. An IT network protocol knowledge is also highly desirable for the candidate to understand network standards, protocols, network design and security.

Third, an understanding of coding and programming languages are critical for a IoT or IIoT professional. Finally, the candidate should understand that change is the only constant and that this arena evolves and changes constantly, so an aptitude for continuous learning will be very helpful.

Positions That Shape the IIoT Industry

An IIoT Solutions Architect plans and creates the technical vision to solve a problem using IIoT. This person designs and manages IIoT solutions and has a start-to-finish responsibility for implementations such as predictive maintenance.

An Industrial Data Scientist works in predictive capabilities for an IIoT system because these devices generate massive amounts of data to be processed and analyzed. Such a position improves the efficiencies of project testing and deployment and develop a comprehensive understanding of the system behavior. The difference between an Industrial Data Scientist and a traditional Data Scientist is that, in the traditional role, analysis is usually performed with static, historical data, whereas an IIoT scientist deals with large sets of data received in real-time.

An IIoT Developer designs, programs, and develops applications (through standard APIs and programming languages) to improve IoT devices. This position creates IIoT applications to lower operating costs, reduce manufacturing cycle time, improve productivity and raise the visibility of manufacturing into the supply chain.

An IoT Embedded Systems Designer plays a key role in developing, testing, and implementing the firmware that enables IIoT devices to function. Programming languages such as C++ and Python are used in the development of device-specific firmware. Candidates are expected to understand computer architectures and hardware security along with computer programming skills.

Evaluating Two Main Players in the IoT World: IBM Watson IoT and Microsoft Azure IoT

The IoT solutions by these two companies provide innovative and integral technologies. Watson IoT is cloud-based and fully managed, which means that it provides centralized dashboards and simple graphical interfaces where users can manage components easily. Key features include APIs to build mobile IIoT apps, and AI-driven analytics and visualization dashboards. For security needs, Watson IoT offers penetration testing, a strong feature in an IoT-cloud environment. Watson IoT has built-in device management as well as scalability for security and provisioning, and its IoT Hub enables over-the-air (OTA) deployment of updates to help maintain IoT devices up-to-date and secure.

Azure IoT is an open edge-to-cloud solution allowing users to operate, from within a single dashboard, all devices and resulting data. Microsoft provides built-in security for every step of the IIoT lifecycle. Users obtain IoT security information and event management, security orchestration, automation and response, and extended detection and response solutions for deployment, devices, edge and cloud services. One key feature by Azure IoT is that it provides customers the option to access an ecosystem of thousands of partners that have already deployed technology solutions for multiple business needs, thus allowing users to streamline IIoT app development for their specific situations. Other key advantages are automation, transformation, and optimization.

SERIALIZING AND TRACKING PRODUCTS

It almost seems like an oxymoron. The more items we produce, the more we want to track them individually. Although individualized tracking of live specifies is also possible through, for example, DNA barcoding, this section will focus on discussing the serialization and tracking of manufactured products.

Individuals, companies of all sizes, government agencies, schools, etc., produce an exorbitant number of products every year (and every year more than the previous one), and many of those products need to be labeled, individually identified, tracked, and traced.

There are several groups of tools at our disposal to serve our needs, as we will see next. It is by no means an all-encompassing list, yet these tools are the most common.

Labeling: labels come in all shapes, sizes, materials, and colors. They can be made of paper, polymers, metal, or electronics. They are printed or stamped with data as simple as a human-readable number or letters, and as complex as QR codes with multiple data sets or hyperlinks. If electronic, the data can either be read-only or dynamic. The industry has created specific printers (thermal, ink-based, laser-etching, etc.). It has also created customized software to create, store, and print the artwork displayed or stored in such labels. Tools and devices to read the barcodes (think linear or 2D scanners or cameras), to interpret and translate codes printed on those labels (such as optical-character recognition tools and devices) have also been developed by the various industries. And finally, specific database management software has been developed to generate, store, validate, generate, and share each individual serial number, model number, and countless other product parameters and specifications, that uniquely identify a product or a set of products.

Barcodes: two types of barcodes are the most used in manufacturing applications. The first type is Linear Barcode (fig 2), also known as One-dimensional Barcode, usually made up of lines and spaces of different lengths and widths. In most cases the barcode also has a human-readable text right next to it (usually underneath it). Some of the most common linear barcodes are known as Codabar and Code 25 (used in libraries), Code 39 and Code 128 (regulated by ISO international standard), EAN 2, 5, and 8 (mostly used for international retail), ITF (for products in bulk), MSI (used in warehousing), Pharmacode (for pharmaceutical packaging), Postnet (for US postal services), and the ubiquitous UPC (for worldwide retail, governed by ISO/IEC).

The second type of barcodes (fig. 3)is known as Matrix Code or 2D Codes. They can represent more data per area than Linear Barcodes. Some of the most used types are Aztec (concentric squares surrounded by vertical and horizontal bars), bCode and BEEtag (for tracking animals), Code 16K (multirow, used for tracking electronic components), Cronto Visual Cryptogram or photoTAN (multi-color,

Figure 2. Linear barcodes

used in e-banking), Data Matrix (for small items, usually electronics), DotCode (pharmaceutical), Han Xin (to encode Chinese characters), MaxiCode (post service, with concentric circles surrounded by dots), PDF417 (postage, airline boarding passes, driver licenses, country visas), the popular QR code (originally designed for automotive components, now used to convey messages through smart phones), SPARQCode (another form or QR Code).

RFID tags: Radio Frequency Identification (Fig 4) is another system to identify and track manufactured products. It consists of the tags (the transponder source) and the reader (a radio receiver/transmitter). Reading is performed much more indirectly and farther in distance than regular barcode readings. RFID uses electromagnetic signals to identify and read tags attached to products.

Figure 3. Matrix codes barcodes

The use of RFID has been extended to track objects such as passports, shoes and clothing, contactless payments (through your smartphone or your credit card), implanted medical devices, library books, inventory tracking, medicines, packages, library books, or even to start your car.

RFID tags can be Passive or Active. Passive tags are limited because they work over shorter distances and transmit smaller amounts of data. They rely on the power of the reader's electromagnetic fields. Active tags are self-powered through embedded batteries; they are able to transmit larger amounts of data over greater distances.

Data transmission from an RFID tag is dependent on the Frequency Band. Passive tags work in low and high frequencies ranging from a few inches to a few feet. Active tags work in ultra-high and microwave frequencies and have larger ranges up to 10 meters. An RFID system relies heavily on radio waves, where electromagnetic fields transmit data from a tag (either embedded or attached to an item) to an RFID reader. The reader uses radio waves to send the signal to the tab and then receive the data back from the tag.

There are potential risks to RFID technology, where the data can be accessed by non-authorized parties that can steal it and exploit it to commit crimes such as financial or medical fraud or identity theft. The process of gathering information from, say, credit cards RFID chips, is known as RFID Skimming. Criminals may insert skimming devices in ATM terminals where the data in your card can be copied to these devices and, eventually, commit fraud such as identify theft. To add protection against RFID skimming, consider using an RFID blocker, such as smart wallets or passport cases that contain protective materials which block radio wave transmission.

Figure 4. RFID technologies

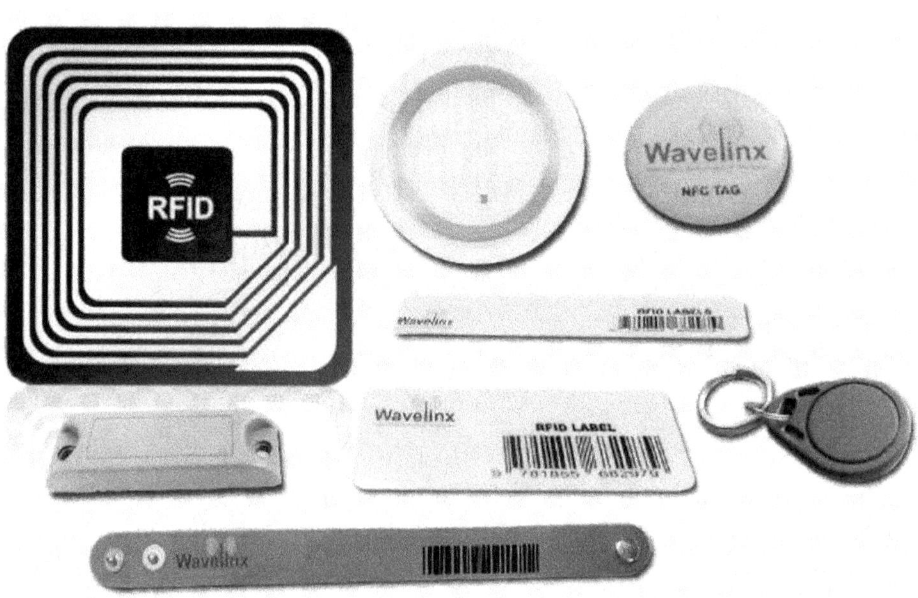

What Are Labels Made of?

The most common types of materials to make labels are: 1) Thermal Paper, used in thermal printers and, in the past, in thermal fax machines; they are composed of high-quality paper coated with a layer of paint (which can be of various colors); they can be waterproof, oil-proof, or friction-resistant, although long-term storage tends to degrade their quality. 2) PVC. Polymer-based, with poor stability to light and heat; they are highly used in electronics, in the automotive industry, and in the chemical industry; their main drawback is the low recyclability properties. 3) Paper-based, usually with pigments or adhesives; their texture is usually smooth and either matte or gloss, with fast ink absorption; they are not water- or oil-proof and moisture is their main foe; their main advantage is lower costs, compared to the other material types. 4) PET (polyethylene terephthalate), a polymer-based material; PET is not as flexible as PVC, but it is more environmentally friendly due to its degradability; its main applications are in the food industry, electronics, and automotive. 5) BOPP (biaxially oriented polypropylene), a film made after stretching it in various directions; it can be mixed-up with other materials such as aluminum film, and PE, to improve performance; its main advantages are transparency, aging resistance, water- and oil-proof, clear printing, luster, and flexibility; usually found in the food, beverage, medicines, and cosmetics.

How Are Labels and RFID Tags Affixed Onto Products?

Labels can be attached to products either manually or through automated label applicators. The automated solution is preferred when speed and consistency is essential, and usually involved specialty labeling software and thermal or laser printers.

RFID tags can be attached or affixed to products in various ways, the most common being: 1) Film Adhesive, the most common method; 2) Foam Adhesive, used where a stronger adhesive method is needed; 3) Epoxy-based, where the strongest adhesive method is needed, such as for tracking tools; 4) with Fasteners, such as screws; 5) attached with Cables or Wires, where the other methods are not applicable; 6) Embedded, to prevent the tag from being removed or its exposition to the environment is either not needed or not desirable.

ERP IN MANUFACTURING

Enterprise Resource Planning (fig 5) is a suite of software that, when applied in Manufacturing (mERP), helps streamline and integrate manufacturing processes across all areas of the operational business. mERP systems help increase an organization’s manufacturing efficiency through management of manufacturing resources, from production floor operations to inventory planning to product and component traceability to product warranties (Gross, J. 2022).

ERP is not inherently in conflict with Lean Manufacturing strategies. Lean Manufacturing systems usually consists of Continuous Improvement (CI) tools such as 5S, Kaizen, SOPs, Lean Office, KPI implementation, preventive/corrective maintenance, Kanban systems, etc. All these tools need and produce data sets that need to be recorded, classified, tracked, monitored, adjusted, and controlled (in a traditional Plan-Do- Check-Act loop), and what better system to manage such data than in a mERP.

An mERP (Sparrow, A. 2022) system will be used, generally speaking, to track a company's manufacturing assets, to serialize finished products and the components (raw materials or subassemblies) necessary to manufacture those products; it can also be used to store the data points and test results as the products are coming down their production lines, to consolidate, report, graph, and monitor those test results and the quantity of products built at any given point in time and by any given section in the manufacturing plant. Customized mERP systems are able to provide product configuration and parametrization capabilities, validation of those parameters for the products being built and tested, and traceability for product warranties, recalls, and returns. With the availability of manufacturing data points in an standardized ERP system, it's relatively easy to generate analytics to continuously monitor a company (or a facility, or a production line, or the weekend shift).

Figure 5. ERP model

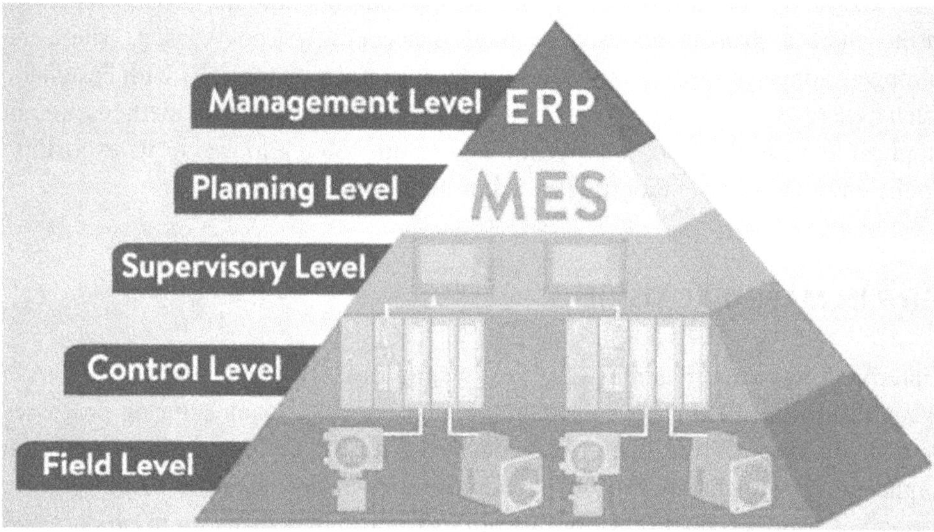

Benefits of mERP

Availability of data points, key metrics, and test results for SPC (statistical process control) and being able toto compare and benchmark with other branches within the same company or with similar companies. This last point will allow a company to evaluate itself within an industry with the goal of becoming a world- class operation. Other key benefits include optimized production operations, minimize product and process variability, improvement of inventory tracking (during warehousing, supply chain, and transportation), real-time inventory management, real-time data sharing, interaction with other company systems, unified metrics (aka KPIs or KPMs), tracking of shop floor activities by assembly station, by product lines, by product cells, or by business unit, tracking of production costs, etc., all with the goal of improving and standardizing production processes with the ultimate goal of customer satisfaction.

Standardized Operating Procedures

SOPs (or Work Instructions), aka Work Instructions, are a key tool in Manufacturing Operations, where products must be built, tested, and released in a standardized, consistent quality all the time. An mERP a perfect tool to design, store, distribute, manage, and make available those procedures.

Inventory Tracking and Replenishment

One of the key functions of an ERP system is to manage a company's supply chain. Either with a pull or a push system, ERP provides standard functionality for companies to leverage its availability. Maintaining product engineering records, inventory levels, replenishment and re-stocking parameters, perpetual inventory records, warehousing management, production planning (when to produce, when to produce, resources needed to produce, etc.), sales and distribution records. All these are activities that are standard to ERP systems in manufacturing environments.

How Is an mERP Hosted?

The principal condition of hosting an mERP system is availability. If the system is down, it's very likely that the manufacturing operations will not function, with resulting negative consequences. Thus, redundancy and operational continuity are key points to consider when determining where and how to host such a system. Three hosting methods are currently available: 1) on-premise, where a company has full control, full ownership, and full support and management of the system; the software may either be internally developed or commercially procured, yet it will reside in the company's premises and under its full control; 2) cloud-hosted, where the software (either internally developed or commercially procured) will reside at a cloud-based host, thus not requiring hardware management by the company, and its availability will not only depend on the host, but on the strength of the connectivity links between the manufacturing facility and the host itself; and 3) hybrid hosting, where the software is deployed both on the company's premises and in the cloud-based host; while providing flexibility in regards to updates and data migration, it may provide functionalities not available on the other two hosting options.

How To Select an mERP System

Several factors should come into play when evaluating mERP system implementations. The main factors are product features, vendor and/or host reputation, cost and support, and ease of implementation. Among the main features to evaluate are: 1) product security, where data availability, data integrity, and data security are paramount; security features should include data backups and restoration, user authentication, data encryption, data accessibility for ease of standard or customized reporting, and data redundancy; 2) integration and interface capabilities, to be able to communicate with other company systems, such as MES (Manufacturing Execution Systems), Human Resource (to manage personnel resources), and reporting services (to be able to provide data analytics and retrieve data for standardized or customized reports); 3) standard industry offerings, which will compare vendor offerings to each other and ensure that system features which are expected to be standard, are a given.

System costs may consist of various models, such as initial license purchase plus ongoing support costs, or subscription model, either per user (be it concurrent or not), or per module, or per location. Another key area to evaluate when selecting an mERP system is the ease or difficulty of implementation. Will your existing personnel be able to implement the system by working directly with the vendor, or will you need vendor's or third-party consultants to help you in the implementation? Will it be sufficient to migrate master data only or will you need to migrate legacy transactional data also? How much of a customization will your implementation require? These are key questions to address, evaluate carefully, and decide before embarking into an mERP implementation.

REFERENCES

Baluch, A. (2023, June 18). Best ERP systems of 2023. *Forbes*. https://www.forbes.com/advisor/business/software/best-erp- systems/

Bishop, E. (2023). What is manufacturing ERP (Enterprise Resource Planning)? *Market Business News*. https://marketbusinessnews.com/what-is-manufacturing-erp-enterprise-resource-planning/289612/

Fernandez, R., Staff, T., Crouse, M., Partida, D., Corrales, E., Ayuya, C., & Kaelin, M. W. (2022, December 6). IOT tech expo global 2022: Exploring Satellite and telco connectivity for businesses. *TechRepublic*. https://www.techrepublic.com/article/iot-tech-expo-global-satellite-telco- connectivity/?utm_source=email&utm_medium=referral&utm_campaign=top-story-of-the-day

Gross, J. (2022). *Leverage ERP to drive lean manufacturing initiatives.* Pemeco Consulting. https://www.pemeco.com/leverage-erp-to-drive-lean-manufacturing-initiatives/

Jabbar, M. A., Tiwari, S., & Ortiz-Rodriguez, F. (2023). Smart Urban Computing Applications. River Publishers.

Johansen, A. G. (2020). *What is RFID and why does it matter?* Norton. https://us.norton.com/blog/id-theft/what-is- rfid

Ortiz-Rodriguez, F., Tiwari, S., Sicilia, M. A., & Nikiforova, A. (2023). Electronic Governance with Emerging Technologies. *First International Conference, EGETC 2022*, Tampico, Mexico.

Sparrow, A. (2022). *Leveraging erp for smarter manufacturing (MOM).* LinkedIn. https://www.linkedin.com/pulse/leveraging-erp-smarter-manufacturing-mom-andrew-sparrow

Chapter 6
Using Blockchain to Decentralize Product Development

Rathishchandra Ramachandra Gatti
https://orcid.org/0000-0002-4086-2778
Sahyadri College of Engineering and Management, India

Chandra Singh
https://orcid.org/0000-0002-1759-8796
Sahyadri College of Engineering and Management, India

ABSTRACT

Digital technologies such as IoT, Cloud, and mobile computing have enabled product innovation, consumer interactions, and efficiency. Blockchain technology has altered how organizations work, providing a paradigm shift in resilience, security, and efficiency. Blockchain enables agile value chains, product innovations, and technological fusion. Every company considers research and product development expensive. Investments are risky, especially if the new product version differs considerably from their current portfolio. Firms must share high investment risks through joint ventures or cooperative product development. However, for collaborative product development, trust and transparency must be vital. Decentralizing design data interchange, funding, and marketing with correct crediting of contributions is one of the methods for multiple-organization collaboration. Blockchains technologies can be employed for such collaborations. This study explores how blockchain can decentralize product development worldwide to meet security, scalability, efficiency, and incentivization criteria.

DOI: 10.4018/978-1-6684-8088-5.ch006

INTRODUCTION

In a typical blockchain, each block is connected to the previous block by a cryptographic hash. The blocks are connected in a chain in which the previous block's hash is stored in the current block's hash, and the current block's hash is stored in the hash of the following block(Singhal, Dhameja, & Panda, 2018). Each block contains private, public, and transactional data (financial and non-financial). Currently, there are two types of blockchain: permissionless or public Blockchain, in which anybody can participate, and permissioned blockchains(Miller, 2019), in which the user needs specific access or permission to participate. Typically, permissionless blockchains are utilized by cryptocurrencies like bitcoin. When data security is of utmost importance, permissioned blockchains are utilized. Several examples in businesses include financial transactions, supply chain networks, health care data administration and marketing data.

Over the years, blockchain has been implemented in numerous areas of management that demand either permissioned or unpermitted distributed ledger-based systems. Blockchains are beneficial in business scenarios involving two or more corporations that wish to do business with each other but do not trust each other or a third party. One such business scenario involves R&D and product development operations(Papakostas, Newell, & Hargaden, 2019) that necessitate substantial unaffordable investments even with the partnership of two or more organizations. With the help of a permissioned blockchain, it is possible to envision a non-trustworthy commercial network among competing organizations.

While developing a permissioned blockchain-based network, desirable qualities such as public verifiability, data transparency, data privacy, data integrity, data redundancy, and trust-anchoring must be met. Depending on the agreement made between the clustering organizations, these attributes may be modified and altered. Such a permissioned blockchain may not be entirely decentralized and requires basic KYC for all new business cluster participants. However, the trust anchor must be specified by the cluster and be able to modify each user's read-write access. This suggests that there is still a centralization of authority, which can be to an elected representative or a board representing the business cluster members' interests.

PROCESS FOR CONVENTIONAL RESEARCH AND PRODUCT DEVELOPMENT

Conventional research and product development processes differ based on the nature of the products, organizational structure, geographical culture of the headquarters, commercial strategy, and external influences. Figure 1 depicts how a typical

Research and Product Development process can be reduced and generalized. It is comprised of basic science research that supports applied science research, which produces a variety of technologies. Both primary and applied scientific research entails substantial investments with meagre direct capital returns and is motivated by curiosity. Consequently, they are suitable for government-funded laboratories or colleges conducting research without stringent deliverables, unlike industrial research, where market demand for research deliverables is significant.

Figure 1. Conventional industrial research and product development process flowchart

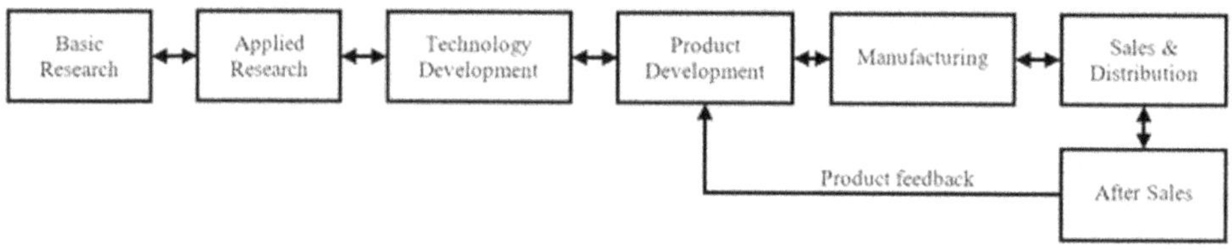

The last phase is product development, which typically involves an industry-driven environment in which one or more technologies derived from primary and applied sciences are combined in a meaningful manner to produce a market-responsive product. Since it is market-driven, additional competitors will attempt to capitalize on the same market with similar products. This chase for market share encourages competition among manufacturers of similar products. Competitiveness motivates the assertion of ownership of technical know-how in the form of various Intellectual property (IP) rights, including Patents, Trade secrets, and Copyrights.

Figure 2. Schematic arrangement of risk-gain for various R&D and PDP phases

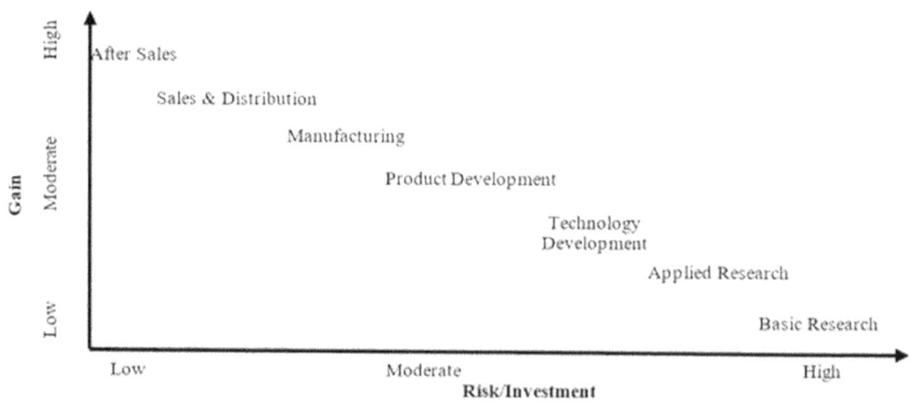

As depicted in Figure 2, the financial risks and rewards associated with the Industrial R&D and product creation processes change progressively from low reward/high investment/risk fundamental research to high reward/low investment sales and distribution.

Although constructing products around technologies may appear to be less expensive than building the technologies that produce those products, it is nonetheless a costly endeavour. This is especially true for significant investment product realizations with uncertain results but high product demand. Data insecurity is an additional problem associated with ownership of product knowledge. Since most machines and processes in industry 4.0 and 5.0 are connected to the internet, they are vulnerable to malicious attacks and hacking(Lee, Abdullah, Jhanjhi, & Kok, 2021), where designs can be readily hacked or reverse-engineered(Corallo, Lazoi, Lezzi, & Pontrandolfo, 2021). Corporate spending on R&D and creating difficult-to-protect intellectual property is becoming a significant concern. These formidable obstacles have compelled competitors to collaborate and decrease financial risks. Apple's collaboration with Samsung for Samsung's 32-bit RISC ARM processor on the iPhone up until the iPhone X, General Motors and Toyota assembling automobiles, and Siemens and Philips developing 4-megabit Dynamic Random Access Memory (DRAM) chips, to name a few, are examples of typical rival collaborations.

If the investment is substantial, it may be argued that product designs can be crowdsourced with the participation of volunteer contributors. This may work for a while, but it also depends on how long people remain intrinsically motivated to keep the project alive and how much money is donated. Maintaining most open-source developers' motivation is essential by using incentives other than credits and donations.

There are typically two-component systems for any product: standard and proprietary. In an automobile, for instance, conventional components include fasteners, starter motors, spark plugs, and brake pads, while proprietary components include the engine, chassis, and gearbox. Now, suppose we wish to enhance the technologies of standard components, such as a material for brake pads. In that case, it should preferably be funded by the auto industry cluster since all automakers will profit. If, on the other hand, the gearbox must be modified for improved technology, it must be proprietary, as it represents the automobile manufacturer's competitive advantage. Therefore, hybrid finance is necessary for a single product in which an industry cluster funds the development of standard components, and the development of proprietary components is self-funded.

New products in emerging regions where standard components are underdeveloped, and customer adoption of new items is limited would be another justification for hybrid finance. Electric vehicles and the electrification of the requisite infrastructure would be one example. Suppose an industrial cluster sponsors components such as

BLDC motors, alternators, and inverters. In that case, they tend to become standard components, and electric original equipment manufacturers (OEMs) can concentrate on limited competitive advantage components such as battery systems and electronic controllers. This would minimize the burden of product development expenses while maintaining market rivalry. Multiple rivals are crucial for preventing oligarchy, wealth concentration, and promoting a healthy market.

When rivals or competitors collaborate, mistrust is always present. This blockchain can be adopted in the cluster to be invalidated. One of the most effective hybrid finance mechanisms for R&D and product development in such clusters is a blockchain that combines the best aspects of permissioned and permissionless blockchains. Thus, there is a clear distinction between what may be published openly and what requires privileged access. The following section considers a hybrid funding model in light of the preceding arguments.

CONCEPTUAL FRAMEWORK FOR THE CYCLE OF INDUSTRIAL RESEARCH AND PRODUCT DEVELOPMENT

As indicated in figure 3, a hybrid finance and IP sharing strategy for R&D and product development is proposed. In contrast to the standard blockchain, where directionality is unidirectional, this architecture employs bidirectional blocks. Similar bilaterality is described in (Papakostas et al., 2019). The bidirectional block B12, for instance, has two blockchain blocks: B12, a block for transactions from node 1 to node 2, and b21, a block from node two back to node 1.

Each bidirectional block symbolizes a step in the R&D and Product Development Process. In contrast, each node represents a station, such as the design office and production floor. Using a bidirectional block, these stations transmit information to the following station.

Each block (Example b12) of the bidirectional block (Example B12) has two hashes - one from the previous node and one from the following node. A block from the design office (node) to the production shop floor (node), for instance, comprises two hashes: the design office's hash and the shop floor's hash. Each block contains both a public key and a private key. The Public key will offer access to all non-confidential information, while the Private key will restrict access to information for people with private access to the permissioned blockchain.

As with other blockchains, tampering with the blockchain or the hash codes destroys the blockchain and invalidates succeeding blocks. It must also be assured that only the data writer of the preceding node has access to write data in the private access data sent on to the subsequent node. The configuration of each block within the industrial R&D and product development process will be discussed.

Figure 3. Schematic of the blockchain conceptual framework for industrial research and product development

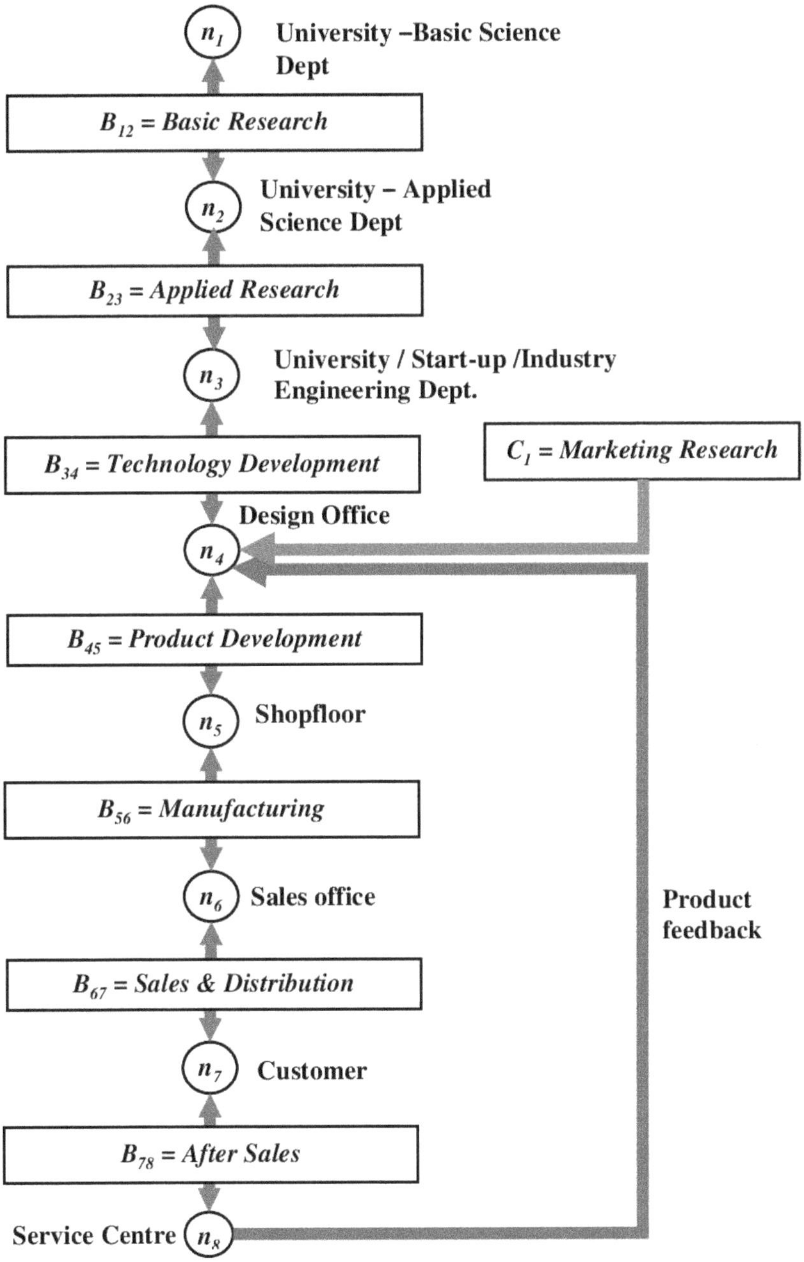

Basic Studies

Curiosity motivates basic research, whose outputs may or may not have monetary worth. Figure 3 depicts it as a high-cost, low-reward procedure, making it generally undesirable to businesses. Consequently, it may be supported by a permissioned blockchain of a consortium of governmental agencies, such as the European Union or a United Nations division for basic science research.

The transaction can alternatively be recorded on an utterly permissionless blockchain of all institutions, where the research index of a university or institute is based on its financial and research contribution to the blockchain. This research index should be connected to university ranking systems (such as QS rankings) and accreditation systems (such as ABET and NBA). In India, for instance, the research index can be linked to the institution's AISHE code, which can then be related to the NIRF rating scheme. It can also be linked to the individual's ORC-ID, where contribution points can be accrued for subsequent career advancement.

Applied Studies

Primary Research outputs serve as inputs for applied sciences research conducted in government laboratories, universities, and to a lesser extent, industrial research centres. Applied research yields technologies that can be used to create various products in the subsequent step, technology development. While there is no capital interest in basic research, applied research outputs (technologies) may provide first-time benefits from which firms wish to profit. Consequently, even if the host organization is primarily a non-profit, it may get funding from corporations under predetermined IP contracts. Consequently, this may have also passed through a permissioned blockchain where write access is granted to faculty members who have entered into predetermined contracts prior to the beginning and execution of the project.

However, the outcome of applied research impacts the entire industrial sector to which it belongs; hence, the industry sector, not the individual, is the primary beneficiary. Therefore, the optimal funders of applied research should be industry clusters. A permissioned blockchain comprises an authorized board for the trust that anchors read-write access based on research contributions (money and data submission). The equity for the collected data and shared bitcoin is proportional to each participant's contribution, which is transparent to the other participants.

A university or institute's research index can depend on its financial and scientific contribution to the blockchain, and transactions can be recorded. This research index must be connected to university ranking systems (like QS) and accreditation systems (ABET, NBA). It can also be linked to the ORC-ID of the individual, allowing contribution points to be accumulated for career advancement.

Technological Progress

The evolution of technology is an iterative process. Once a technology is created, it is periodically enhanced with new characteristics frequently derived from other technologies. Typically, technology exists at the component or subsystem level. Since most original equipment manufacturers (OEMs) outsource their components, technology advancement depends on big suppliers. If numerous small-cap companies utilize a particular technology, it is unlikely to receive R&D and PDP funding. In such situations, a permissionless or permissioned blockchain with diluted permissions will be the optimal adoption to speed up research. If large-cap hardware manufacturers dominate the market for the specific technology, then a permissioned blockchain is appropriate.

Product Development

The companies' financial interests begin with the product development phase and continue through the after-sales period. While research and development of new technologies fuel the creation of new products, their findings cannot be directly translated into marketable products. In contrast, the output of the product development phase is saleable products that are of interest to businesses.

There are currently two product development methodologies: proprietary and open source. While producing a proprietary product requires high data protection for intellectual property, developing an open-source product entails no data security. Therefore, limited users can share personal items via permissioned blockchains, but open-source products can utilize permissionless blockchains.

In reality, however, goods are comprised of standard component systems that are available to the public and proprietary components that are kept confidential within the organization. Consequently, a flexible hybrid approach utilizing a hybrid permissioned blockchain is required where the trust anchor can adjust the access control as necessary. By offering access, even outsider experts, such as university professors who do not belong to the company, can assist in product development. However, data security will remain a worry with permissioned blockchains.

Manufacturing

Throughout the design for manufacture, prototype, and pilot run phases of a product development process, manufacturing must collaborate with the design office to ensure that the product is optimized for production based on the available manufacturing infrastructure. Since most of the machines in Industry 5.0 are connected to the internet, the integration of manufacturing nodes from outside the organization will be relatively seamless. Machine-to-Machine (M2M) communications connect digitally connected machines for data exchange. In manufacturing, digitally connected equipment facilitates the implementation of permissioned blockchain.

M2M communications can be restricted through permissioned Blockchains, wherein the supplier of CAD or design services can upload their engineering data but cannot execute the command on the CNCs or the machines to perform machining operations. The machine control operations are purely offline and cannot be accessed through internet to the other users.

Permissioned blockchains permit the secure integration of both prototype and product design at the level of component systems, hence accelerating the product development process. It also enables supplier-level prototyping during product development, lowering manufacturing risks associated with outsourced component system designs. Further data control is possible in the product data management by not disclosing the features in CAD files (for example sharing only the IGES or STL file to the supplier) making it difficult to reverse engineer (although not impossible).

Material data files and the product specifications can be controlled at field level with only the selected fields to be visible to the external stakeholders. For example, only the mechanical properties of a particular material to be outsourced can be permissioned to be shared to the bidding suppliers with no access to the material composition.

The supply chain can also be controlled at country level wherein the product components are outsourced to different countries and the final assembly can be kept proprietary by varying the degrees of permissioned fields for suppliers of different countries. This ensures data integrity of the current and future product design data of multinational companies having global supply chains.

Sales and After-Sales

Through permissioned blockchain, sales and distribution may be networked, with GM-Marketing establishing trust with the local sales manager and granting access to sales personnel and distributors. For product development, product feedback such as product ideas (during sales), complaints and warranty claims (during after-sales) can be periodically collected in CRM and used by the design office to improve future product variants.

The Marketing data needs to be highly secure and purely proprietary thus keeping this data not accessible to rivals. However, the data can be shared through permissioned Blockchain to the suppliers to improve the supply chain quality and processes and thus improved delivery. However, marketing data needs to be highly regulated in permissioned Blockchain if there is threat of forward integration from the suppliers.

The standard spare parts drawings can be made permission less so that they can be manufactured by the suppliers. Only the critical parts which is high cost, low volume parts and add significant revenue to the after sales can be controlled using permissioned Blockchain. Thus the after sales design data, warranty claims and other information can be carefully controlled using permission variability in Blockchain technologies.

CONCLUSION

Permissioned and permissionless blockchains each have their advantages and disadvantages. Both can be utilized in industrial research and product development, but only in certain circumstances, to speed up and reduce product development expenses. Employing blockchain enables expanded engagement from both within the firm and external stakeholders such as suppliers, customers, and even rivals. Heavy investment risks can be mitigated when mutually mistrustful, competitive organizations interact, which would not be achievable if facilitated by a third party. Employing blockchains for R&D funding and PDP funding eliminates ambiguity and bias, increasing participation. In addition, organizations do not have to worry about the decentralization of data, as permissioned blockchain enables the decentralization and retention of confidential data. This study is a preliminary investigation into the application of blockchain in industrial R&D and product development. The ongoing study focuses on the implementation phase.

In the implementation of this controlled shareable product design framework, any company can adopt both open sourced ideas and proprietary ideas in their product design and manufacturing through controls in Blockchain networks. The degree of proprietorship can also be varied on a product by product basis. Industry standards for the products of the competing companies can be co-invested by the competitors to reduce costs. The redundancy of research and product development costs and efforts can be minimized by shared resources through Blockchain. Also, the product designs can be realised very soon with reduced design cycle time thereby accelerating the time to market and reduction in opportunity cost losses.

REFERENCES

Corallo, A., Lazoi, M., Lezzi, M., & Pontrandolfo, P. (2021). Cybersecurity challenges for manufacturing systems 4.0: Assessment of the business impact level. *IEEE Transactions on Engineering Management.*

Lee, S., Abdullah, A., Jhanjhi, N., & Kok, S. (2021). Classification of botnet attacks in IoT smart factory using honeypot combined with machine learning. *PeerJ. Computer Science*, *7*, e350. doi:10.7717/peerj-cs.350 PMID:33817000

Miller, A. (2019). Permissioned and permissionless blockchains. *Blockchain for Distributed Systems Security*, 193–204.

Papakostas, N., Newell, A., & Hargaden, V. (2019). A novel paradigm for managing the product development process utilising blockchain technology principles. *CIRP Annals*, *68*(1), 137–140. doi:10.1016/j.cirp.2019.04.039

Singhal, B., Dhameja, G., & Panda, P. S. (2018). How blockchain works. In *Beginning blockchain* (pp. 31–148). Springer. doi:10.1007/978-1-4842-3444-0_2

Chapter 7
Why Is Manufacturing Cybersecurity So Challenging?

Alexei Pinal-Avila
Nozomi Networks, Mexico

ABSTRACT

Cyber threats against industrial manufacturing are escalating as more and more connected systems create larger attack surfaces. Being that cyber-attacks are one of the top concerns to interrupt operational continuity, it is critical for manufacturers to establish a complete program to protect, identify, detect, respond, and recover. Highly accurate detection of cyber threats, risks, and anomalies improves operational resilience, no matter the age of the facility or the scale of the production lines. In terms of industrial cybersecurity, there are specific challenges to take care of: firstly, Legacy manufacturing devices were designed for reliability, not security; secondly, network segmentation and zero trust have been slow to come to manufacturing; and exponential growth of IoT sensors.

INTRODUCTION

Cyber threats against industrial manufacturing are escalating as more and more connected systems create larger attack surfaces. Being Cyber-attacks one of the top concerns to interrupt operational continuity, it is critical for manufacturing to stablish a complete program to protect, to identify, to detect, to respond and to recover. A highly accurate detection of cyber threats, risks and anomalies improves operational resilience, no matter the age of the facility or scale of the production lines.

DOI: 10.4018/978-1-6684-8088-5.ch007

In terms of industrial cybersecurity, there are specific challenges to take care: firstly, legacy Manufacturing Devices were designed for reliability, not security; secondly, Network Segmentation and Zero Trust have been slow to come to manufacturing industrial networks; thirdly, exponential growth of Internet of Things (IoT) Sensors and as last an economic challenge: budget assignment by management for a cybersecurity program.

MAIN CHALLENGES OF INDUSTRIAL CYBERSECURITY FOR MANUFACTURING

Firstly, most of manufacturing facilities usually have Industrial Control Systems which are designed for decades-long lifecycles and designed for reliability rather than security. Operational Technology (OT) systems also often have a higher lifetime of 15–20 years, or more, compared to 3–5 years in the Information Technology (IT) industry. Usually, they often run minimal versions of operating systems and communicate using proprietary or specific industrial protocols that lack security features as encryption or authentication.

Secondly, Operation's and Maintenance's engineers are reluctant to deploy traditional IT cybersecurity techniques such as network segmentation or zero trust solutions. A misconception of adding complexity and overhead to cyber physical process management and introducing risks to production uptime and safety prevent a rapid adoption of such technologies. Industrial control systems have less tolerance to downtime and are responsible for controlling physical processes. In contrast to IT where Confidentiality, Integrity, and Availability (CIA), Safety and Reliability are the most important drivers in daily operation of lines of production.

As third challenge is that millions of Internet of Things (IoT) devices are being added by daily basis to industrial manufacturing networks to reduce costs and deliver more value to customers: fast response, safety, reliability, sustainability, etc. Like legacy Industrial Control devices, many IoT devices (Jabbar et al 2023) have lightweight operating systems that lack common security features. Also, their firmware is rarely updated, and they are frequently connected to other systems and provide an easy attack vector for hackers. IoT devices generate extensive data that requires extensive processing to achieve adequate learning for being protected, but relying on local learning by each IoT device is not feasible in most cases due to its mentioned limited resources of memory and processor's capability.

Basically, a current manufacturing factory is nowadays a highly digitized (Ortiz-Rodriguez et al, 2023) production facility that is based on fully integrated, collaborative manufacturing real time systems and automatic response to dynamic conditions. Additionally, the current revolution of IoT sensors is improving notably the efficiency and effectiveness. But Covid19 pandemic exposed global supply chain weaknesses and industry vulnerabilities; factories realize that current manufacturing systems are full of legacy systems without intrinsic cybersecurity and worst without any further consideration of access remote control, malware detection or security incident logging. For instances, there will be presented some typical attack scenarios in OT environment: compromise of Human Machine Interface (HMI) or Operator Stations (OS) or Engineering Stations (ES); modification of process or controller parameters; and unauthorized connection detected or unauthorized device connected.

First case, the compromise of Human Machine Interface (HMI) or Operator Stations (OS) or Engineering Stations (ES). Here a malicious actor may cause a denial of view to disrupt and prevent operator oversight on the status of an OT environment. The awareness of change of process variable can be temporally block and prevent proper operations. The loss of view will require local, hands-on operator intervention, for instance, a restart or manual operation of a pump or a robot. A Second case, a Program Logic Controllers (PLC) perform actions based on process variables, the values are stored as parameters on memory. Such parameters can determine the extent to which an action is performed. Modification of these parameters can produce an undesired outcome, for instance, a non-planned shutdown of a motors who manages a conveyor.

As third scenario, (Sandeep Lota, 2023)a malicious actor may also setup a non authorized communications among servers and PLCs. The imposter server can be used to send legitimate control messages to other control system devices. These impersonated control telegrams may be turned into dangerous, out-of-bounds, or unexpected values from typical operations. An unauthorized command messages to instruct control system assets to perform actions outside of their intended functionality. The bad actor may cause disruption of the manufacturing process or destruction of manufacturing line of production. Special case are the attacks over Real Time Locating System (RTLS) that track materials and products inside the lines of assembly. A malicious actor may connect an unauthorized device to wireless networks and then compromise of a wireless device and this rogue device As last case a non-authorized actor could gain access to the enterprise network and use this as initial access into the OT environment giving access to the servers were all data historian with all registers of quality systems or intellectual property are safeguarded. Access to data historian can be used to exfiltrate its data that can be used to learn about the process, control systems, and operational details. All competitive advantages could be lost, additionally that this access can be subsequently used to launch further attacks into the OT systems.

The king Key Performance Indicator (KPI) in manufacturing is Overall Equipment Effectiveness (OEE), then it comes up quality cost reduction, inventory reduction, speed to market time reduction, operating cost reduction and energy efficiency. All costs were allocated in being more effective and efficient manufacturer, but one important pillar has been disregarded: keeping cyber safe all previous investments. But most manufacturers start reporting an exponential growth of security incidents as phishing or ransomware. Being Ransomware the greatest threat to manufacturers because it can shut down production and disrupt day-to-day operations. As factories are digitized, their attack surface is exponentially increasing, requiring security visibility into all IT, OT and IoT systems across all facilities. So then operational assets in industrial networks must be identified, monitored, and secured. Asset Inventory and Vulnerability Management are the first step to reach cyber resilience.

The statement "You cannot protect what you cannot see" continues being more valid than ever in OT and IoT environment. Visibility enhances automates asset inventory creation, eliminating blind spots and revealing assets others don't see. Visibility must provide accurate and extensive information on all OT, IoT and IT assets from all system, no matter their age, vendor, or function. Automatic tools will enhance from time-consuming manual data collection and spreadsheet work to accurate and immediate usable dashboards. Increasing vulnerability visibility reduces the mean time to detect (MTTD), thereby having a better situational awareness. In Operational technology Visibility without prioritization is not worthy. A comprehensive, up-to-date inventory of OT cyber assets is not complete if categorization of the associated physical process elements is missed. An easy step is to define critical assets without which lines of production cannot operate. Then, the assets that are used at least once per day and can disrupt normal operations and then assets which are used less frequently than once per day.

On one hand, digital transformation is the only path towards maintaining a competitive advantage. On the other hand, it makes factories more vulnerable to cyber risks and downtime threats. Companies with a mix of brownfield and greenfield sites need a solution that reduces cyber risk and improves operational resilience. Improving operational reliability requires real-time situational awareness. By learning factory's normal network patterns and manufacturing processes, an automatic solution that identifies anomalies or cyber intrusions will help OT engineers to act before equipment failure, unusual variable values, or network stability issues impact production uptime or quality. Malicious actor encompasses from a malicious insider to an advanced persistent threat (APT) actor. Both bad actors could put on risk the operational continuity.

Unplanned downtime can occur for multiple reasons: hardware breaks down; a change in the industrial network impacts the production lines; or a cyber incident disrupts communication or process execution. While OT engineers are figuring out the problem and how to be troubleshooting, OEE is hit and, operating cost increases, in simple words, the factory is losing revenue due to the downtime. Being informed of cyberattacks, vulnerabilities and risks to safety and reliability metrics in accurate manner will pinpoint alerts and actionable insights to help OT engineers to mitigate cyber and operational threats before they cause harm to operators and damages to the machinery. Increasing anomalies and intrusion's visibility reduces the mean time to respond (MTTR), thereby improving operational resilience. Overall, all data needs to be integrated seamlessly with Security Incident Event Managers (SIEMs), asset management systems (AMS), firewalls (NGFW), and ticketing systems. A holistic integration improves cyber resilience notably.

It is extreme difficult to cover all attack surface due to following factors: exponential growth of attack surface and the millions line of code on each Industrial control system software platform. A software platform is composed by thousands of code sources. Most Code programmers focus on functionality more than cybersecurity. Therefore, it is mandatory the identification of top adversarial techniques, awareness of cooccurrence rules among adversarial techniques that can help cyber analysts in threat hunting.

Manufacturers need process and stability monitoring for automation systems of production. The monitoring gives unusual variable values (Davide Boglioli, 2023), early warning of failing equipment, and networking communication anomalies. Detection cyber incidents prior to impact should be part of a holistic security architecture. Detection capability should cover initial access, discovery, and lateral movement. Cyber attackers can breach defense tools through multiple attack vectors, and thus defending against attacks requires an understanding of how adversaries deploy techniques. The use of frameworks as MITRE ATT & CK allows the analysts to identify specific adversarial techniques in cyberattacks and spread technical reports how the menaces operate.

CYBERSECURITY MAIN ELEMENTS

However, a resilient defense should (Mc Carty et al. 2023) also be prepared to restore and recover when a cyber security incident is not found until it is too late. Once a cybersecurity event is detected, the tasks that should take place before the event is satisfactorily solved are event reporting, logging review, event analysis, incident handling and response, eradication and finally recovery (Michael Powell et al. 2022).

1. **Event Reporting**. Once an event is detected, it should be reported to the appropriate predetermined stakeholders for initial triage. The triage process will assign an appropriate priority for handling the incident.
2. **Log Review**. Keeping records of event/audit logs provides support for forensics, which allows identification of root causes, technical vulnerabilities, behavioral vulnerabilities, and improvement opportunities.

3. **Event Analysis**. The security-related events should be analyzed to identify and characterize attacks, security compromises, and security incidents.
4. **Incident Handling and Response**. A process for evaluating cyber incidents should be used that identifies the potential impacts, threats, and vulnerabilities that allowed the incident to occur. An appropriate response should include containment, reducing the impact, applying counter measures to mitigate root cause, and protecting the OT against future threats. Assignment of persons in charge for incident and response and disaster recovery processes, with their names and contact detail is essential, so they can follow ongoing backups, maintain operational continuity systems, declare of an incident and issue a report to management and handle relevant stakeholders as customers, press, etc.
5. **Eradication**. Once the incident is contained, all means of persistent access into the network should be eradicated so that any malicious actor activity is sufficiently limited. Typical tasks are implementing safe perimeter, firewall rulesets and boundary router access control lists. Then OT engineers reconnect the rebuilt systems to the network, restore systems to normal operation, monitor operations for abnormal behaviors and backup newest configuration.
6. **Recovery**. The goal is to allow return of normal operations by eliminating attack vectors of the incident and mitigating the vulnerabilities or other conditions that were exploited. Some typical actions are remediating all infected systems in the OT environments; identifying and validating the integrity of correct backup before restoration; reformat affected systems or rebuild systems from scratch; installing security patches; resetting account passwords, replacing compromised files with clean versions, and downloading safe versions of the PLC and HMI programs. Recovery plans might look great on paper but fail when they are needed most because none has practiced them. To avoid this from happening, OT engineers must run drills and test the recovery plan in a realistic scenario. Learning the lessons from the drill and updating the plan of disaster recovery, so it is clear and more effective for all parties involved. It is essential to include main metrics for this phase: Recovery Time Objective (RTO) and Recovery Point Objective (RPO). A disaster recovery plan must make it clear what your factory is expecting: RTO as the maximal time the factory can tolerate for recovering normal operations in case of a disaster (for example, recovery within 60 minutes, 2 hours, 12 hours, etc.); RPO as the maximal amount of Production your organization can afford to lose (for example, an hour of Production, 3 hours of production, one day of production).

Cyber controls that the factory can put in place to alleviate the risk from one or more attacks by reducing the probability of these attacks successfully exploiting vulnerabilities An OT Engineer chooses to implement a control at a certain level for the manufacturing center. Two declarations are valid: The higher the level of desired specific cyber security control the greater the degree to which the control is implemented. Second the wider the level of diverse cyber security tools the higher the degree of cyber resilience. An implementation of level has a degree of vulnerability mitigation on each target and execution of 2 or more controls covering more attack surface causes reduction of probability of occurrence.

The cybersecurity technologies (S. Dambra et al, 2020) found to be minimum set of tools to reach cyber resilience are:

1. Operational technology encryption that provides confidentiality to industrial communications by preventing communications from being intercepted. One of the downsides of encrypting is a loss of deep packet inspection visibility. A malicious command may be sent over an encrypted channel, and it is not possible for Network-based Intrusion Detection (IDS) to detect that operation and send an alert.
2. Then Network-based Intrusion Detection (IDS) uses network data to identify anomalous communications and alert. An IDS requires a training period to learn normal network patterns. This training data set needs to include a full range of operations or there is the risk of false positives. There is a point of concern the likelihood of detecting some malicious long extended actions, it decreases because the IDS begins to recognize this traffic as normal.
3. Endpoint Detection and Response (EDR). It identifies anomalous actions on endpoints (user logons, account creation and modification, binary execution, etc.). It is essential that Operation Stations and Engineering stations should be carefully configured to ensure logging information is available.
4. After detection, correlational analysis is the next step. Security Information and Event Management (SIEM) systems will add information from different sources of security monitoring and detection tools (EDR and IDS), threat feeds, etc., and potentially integrated analytics for detailed analysis and correlation. Extended Detection and Response (XDR) and Security Orchestration, Automation, and Response (SOAR), such technologies ingest IDS and EDR data and use automated response tools to defend the system assets with automated playbooks that provide a step-by-step flow of actions to expediting response.

OT CIBERSECURITY

The information of cyber security is presented in different ways for the OT engineers responsible for the secure, reliable operation of the manufacturing centers than the Chief Information Officer (CIO) and Management. Engineering department requires a technical correlation of the benefits to ensure the security and reliability. But C-Leaders require correlation of the benefits in terms of savings from cybersecurity insurance expenses, probability of losses, etc. CISO (Chief Information Security Officer) in collaboration with manufacturing managers should plan over where and what level of implementation degree (Andrew Fielder, 2016). Industrial control systems in Manufacturing are being attractive targets for cyber attackers because of constant revenue and the urgency of being up and running. However, the factories are heavily restricted with the available funding for cyber security, generally working with a fixed budget with little to no additional funding being made available for cyber security purposes. It is generally perceived that this budget is insufficient for them to cover all the vulnerabilities that the system may have.

Industrial organizations must consider two critical factors; the cost of implementing a particular defence and the impact that defence has on the production line. The first one considers only the direct cost of implement defense within limited budget. The second one considered the indirect cost of the defence. It means the consequences of the cyberattack. Examples of indirect cost are System performance cost, morale cost, and retraining cost. The sum of costs is the sum of the security damages inflicted by a cyberattack and the indirect cost for implementing a control at a certain level.

Direct cost is divided into two categories, the Capital cost and Labor cost. Each manufacturing organization is likely to have different grade of automation systems with diverse sizes, age of systems, and configurations; in terms of a direct financial value, it is difficult to calculate a monetary value. An over specialized budget requirement would make an implementation infeasible in the real world. A low general budget requirement would make the cyber defense weak and limited. The way to solve this is to normalize the direct costs of the control in timing tasks for instance, weekly patching. This is something measurable and achievable.

Indirect cost is as an expected level of additional disruption caused in one of three categories: System Performance, any reduction in the speed and capability of the system to perform the related process tasks; User morale, the impact of the control on the behavior of the automation users; and retraining that is the additional requirements to be able to use the new cyber security control. First one, it may be given by direct amount of non-produced articles in a period. User morale is intangible and foreseen effects are in middle and long term: attrition, toxic work environment, and short skilled workforce.

CONCLUSION

This chapter has significantly extended the importance of industrial cybersecurity in the sector manufacturing. It summarizes the selection of security controls and steps in case of cyber incidents. It also highlights how cyber security investment should be generally analyzed. In a nutshell, this chapter explained a general understanding of cyber security on the line of production for manufacturing. The attacks to Operational Technology are exponentially growing and each day are more sophisticated. A compressive and holistic approach of industrial cybersecurity by all management and workers are the only to road for being cyber resilience.

REFERENCES

Boglioli, D. (2023). *The Challenges of the Vulnerability Assessment Process in ICS Space*. Nozomi Networks. https://www.nozominetworks.com/blog/the-challenges-of-ics-vulnerability-assessment-process/

Boglioli, D. (2023). *EBOOK-MFG-Improving-Op-Res-8.5x11-001.* Nozomi Networks. https://www.nozominetworks.com/downloads/US/Nozomi-Networks-Manufacturing-Improving-Operational-Resilience-eBook.pdf

Dambra, S., Bilge, L., & Balzarotti, D. (2020). SoK: Cyber Insurance – Technical Challenges and a System Security Roadmap. *2020 IEEE Symposium on Security and Privacy (SP)*, San Francisco, CA, USA. 10.1109/SP40000.2020.00019

Fielder, A., Panaousis, E., Malacaria, P., Hankin, C., & Smeraldi, F. (2016). Decision support approaches for cyber security investment. *Decision Support Systems, 86*, 13-23. doi:10.1016/j.dss.2016.02.012

Jabbar, M. A. (2023). Applied Machine Learning and Data Analytics. *5th International Conference, AMLDA 2022*. Springer Nature.

Lota, S. (2023). *A Comprehensive Guide to IoT Security*. Nozomi Network. https://www.nozominetworks.com/blog/guide-to-iot-security/

Mc Carty, M., Johnson, J., Richardson, B., Rieger, C., Cooley, R., Gentle, J., Rothwell, B., Phillips, T., Novak, B., Culler, M., & Wright, B. (2023). Cybersecurity Resilience Demonstration for Wind Energy Sites in Co-Simulation Environment. *IEEE Access : Practical Innovations, Open Solutions*, *11*, 15297–15313. doi:10.1109/ACCESS.2023.3244778

Ortiz-Rodriguez, F., Tiwari, S., Sicilia, M., & Nikiforova, A. (2023). *Electronic Governance with Emerging Technologies*. First International Conference, EGETC 2022, Tampico, Mexico.

Chapter 8
Bibliometrics Study of Organizational Cybersecurity

Francisco Isaí Morales-Sáenz
https://orcid.org/0000-0001-9740-149X
Tamaulipas Autonomous University, Mexico

Jose-Melchor Medina-Quintero
Tamaulipas Autonomous University, Mexico

Fernando Ortiz-Rodriguez
https://orcid.org/0000-0003-2084-3462
Tamaulipas Autonomous University, Mexico

ABSTRACT

Cybersecurity is one of the main concerns of most organizations, derived from the violation of their cybersecurity. The objective of this chapter is the development of a bibliometric study on cybersecurity from the organizational level. For the method, the Web of Science platform was chosen, since it indexes and organizes scientific production and has the high impact factor of the Journal Citation Report (JCR). The results show evidence of the evolution in which cybersecurity has been addressed within the organizational field, so the originality is related to the business importance in this globalized world. The various theories and approaches used in empirical studies provide valuable perspectives to understand the behavior of the actors involved, the motivations behind criminal actions, and cyber risk prevention and mitigation strategies, contributing to the good performance of organizations. These findings show the interest of the scientific community, observing a growing trend around publications within the subject of study.

DOI: 10.4018/978-1-6684-8088-5.ch008

INTRODUCTION

Today, the development of digital technologies is having an impact on all sectors of society (Stetsenko et al., 2021), by providing benefits for companies and even governments allowing them to increase the speed and accessibility of information communication technologies through better performance, productivity and reduction of operating costs (Yong et al., 2010; Park et al., 2017). Therefore, information technologies play a fundamental role in the development of business activities, as a source of competitive advantage, since organizations have access to a better position in the market (Hernández & Salcedo, 2020), derived from the wide spectrum of business opportunities that trigger the general strategy within companies (Latilla et al., 2021).

However, the benefits provided by information technologies, their implementation and use in productive activities often entail serious managerial challenges that impact information security (Hasan et al., 2021); in this sense, it has been pointed out that cyber threats that violate systems within organizations are becoming more frequent as a result of the ease in preparing and executing this type of threat. The foregoing has caused an increase in criminal activity within the cybernetic area, for which there are no borders, attacking national, industrial and financial interests (Dzyana et al., 2022).

One of the phenomena that demonstrated the total dependence on the technological field for most sectors of society was the SARS-CoV-2 pandemic, which led to the search for and implementation of technological solutions for adaptation to the new normality and, brought with it the need for an adequate level of cybersecurity to protect the remote workforce as well as the technological infrastructure of organizations (Furnell et al., 2021), as cyber-attacks focus on vulnerable people and technological systems (Pranggono & Arabic, 2020).

This global phenomenon triggered a series of unprecedented information security concerns, derived from a worldwide increase in scams and cyber fraud, inducing the search for solutions and strategies to deal with this type of attack worldwide (Al-Qahtani & Cresci, 2022), for governments, society and, above all, for organizations (Carrapico & Farrand, 2020), who had a massive migration from face-to-face professional activities to an online environment, which has raised a problem new and context for cybercrime (Jafar et al., 2022); in which organizations that are not prepared to face cyber-attacks on their organizational resources will experience huge losses and, above all, will have a negative impact on their performance (Berlilana et al., 2021).

It is estimated that cybersecurity threats will cost the world around 10.5 trillion dollars annually by 2025 (Morgan, 2020), however, the actual costs could be higher because some organizations may avoid providing reports on threats or attacks received for fear of generating a bad reputation that affects their image (Kim & Lee, 2021). Among the costs of cybercrime are: damage to technological infrastructure, loss of productivity, theft (money, intellectual property, trade secrets, customer data), fraud, as well as damage to the organization's reputation (Ahmad et al., 2021; Hu et al., 2021; FBI,2021).

Derived from the relevance of computer security for organizations, it is extremely important to search for alternatives, solutions, as well as strategies that protect their assets, both tangible and intangible, related to information technologies, in which the information technology can be prepared organization to anticipate attacks on its computer systems and networks that could reduce the performance of business operations.

On the other hand, one of the most used tools to determine research trends on a given topic is the bibliometric study, which identifies research trends and obtains the most outstanding aspects and research contributions (Van Eck & Waltman, 2019). Regarding the articles that investigate the aspect of cybersecurity from a bibliometric perspective, there are studies such as the one by Loan et al. (2022) in the period from 2011 to 2020, their findings show an increasing annual trend within the field of scientific research, an interesting aspect to observe are some perspectives on which the study is addressed, such as the contribution by country, the authorship, trending topics, collaborations and network mapping on the literary production of cybersecurity.

However, there is a gap within the field of scientific research around cybersecurity seen from an organizational perspective, which allows us to get closer to knowing how this aspect has been studied, as well as the different perspectives and contributions that have been made as a contribution to the knowledge.

In this sense, the purpose of this research is to study the state of the art related to cybersecurity in organizations, specifically using the Web of Science (WoS) platform with indexing in the Journal Citation Reports (JCR), in order to analyze the articles published within this platform. The foregoing will promote the generation of a bibliometric report that supports the understanding of the subject of study and the development of research in the area over the years, and in this way, related information can be obtained on some basic aspects of bibliometrics that answer the following questions:

- What has been the growth trend of scientific production related to the study of cybersecurity from an organizational perspective?
- Which researchers have made the most contributions on the subject and which are the most cited?
- Which are the most relevant publications regarding the number of citations within the scope of study?

- What theories have been used in the study of the topic?
- Which countries is where the study of cybersecurity has been most developed?
- Which universities lead scientific production in this type of research?
- What are the key terms used by researchers for the analysis of this topic?

The answer to these questions may contribute to the field of study and the information may be relevant for mapping and clarifying organizational cybersecurity and be used by cybersecurity practitioners, professionals and researchers, to determine in a limited and summarized way the relevant fields within the investigation, have a guide on what has been done and the trend in this particular topic.

To achieve this, the document is structured as follows: the method section describes the process carried out to achieve the objective and the steps followed to carry out the bibliometric analysis. In the results section, the analysis of behavior and its evolution over the years is presented and, finally, in the conclusions, the findings are discussed from a general perspective, as well as the limitations and future lines of research.

METHOD

In this section, the bases are established regarding the tools and materials implemented to achieve the proposed objective. To do this, the database under analysis is defined, as well as the terminology and search criteria for the documents, then the process of the tool applied for data analysis is described, as well as the one used to generate the mapping of the data networks of scientific production.

In this sense, the selected unit of analysis is Web of Science, due to the fact that its importance at the international level has been highlighted, making it one of the most popular and recognized platforms within the scientific community, in addition, an indicator is integrated that determines the impact factor, such as the Journal Citation Report (JCR) (Goyanes & Demeter, 2020), so the export of metadata from said platform allows the execution of bibliometric analyzes on a certain field of study (Kipper et al., 2019).

The literature review is one of the most relevant approaches within the area of scientific research, it is the origin of the mapping of theoretical themes, as well as the identification of knowledge gaps within the literature (Snyder, 2019). Therefore, the search for primary information sources is carried out, oriented towards scientific production related to the organizational cybersecurity terms indexed within the platform.

The criteria used to search for publications within the Web of Science is: *Cybersecurity AND Organization OR Organizational Cybersecurity*, broadening the spectrum of study to all areas of knowledge within the database, reaching a total of 345 publications, ranging from 2006 to 2022. Said information was exported in. bib format, for processing in the statistical software *R*, specifically with the use of the *bibiometrix* package, which is frequently used within the study of scientometrics to obtain efficient indicators bibliometrics, which in the case of this research, will allow the objective to be achieved (Aria & Cuccurullo, 2017).

The indicators that were used for the study are the annual scientific production, the most relevant journals, the researchers who have contributed the most to the generation of knowledge on the subject, the most relevant documents by number of citations, and the most used terms in this particular field. In addition to the above, the mapping of contribution networks by countries is used, as well as the links around the terminology used with the help of the *VOSviewer* software, which is a tool widely applied within bibliometric studies when using maps of networks with the purpose of showing the structure and interrelationships between different elements such as authors, journals, universities or countries (Van Eck & Waltman, 2010), and synthesizes the characteristics of the state of the art on a certain subject, offers an alternative for the understanding and analysis of research trends through their implementation (Hallinger & Kovačević, 2019).

RESULTS

The execution of the analyses, in which the period of the first publication dates from the year 2006 to the year 2022, with a total of 345 documents; According to its type, the distribution is 305 articles, 13 early access, 2 proceeding papers, 1 book review, 1 editorial document, as well as 24 review documents, which together cover a total of 1398 keywords and 1058 authors, of which only 51 appear as sole author and 1007 as co-authors of the documents analyzed.

The languages of the publications are mostly English, with a total of 333 documents, 5 in Russian, 2 in Portuguese, 2 in Spanish, complemented by French, Polish and Ukrainian with 1 respectively.

Regarding the categories and research areas in which they are grouped, there are computer sciences, artificial intelligence, engineering, economics and business, law and government, science and telecommunications technology, as well as within the field of social sciences. In relation to the publishers that publish on the subject, there are Elsevier, MDPI, Springer, IEEE, Emerald Group Publishing, Wiley, SAGE, Taylor & Francis, Oxford University Press and IGI Global.

After this brief analysis related to general information and with the purpose of achieving the proposed objective, the analysis of the bibliometric indicators continues, beginning with the review of the annual scientific production, the most relevant journals, the impact and scope of the journal within the platform, authors, universities, scientific production by country, the most relevant documents, the most used concepts, the mapping of collaboration networks between countries as well as the key terms most used in publications within the area of interest.

Annual Scientific Production

Regarding the scientific publications within the Web of Science database, Figure 1 indicates the production, covering a period of time from 2006-2022. A constant increase is shown from the year 2006, showing a constant trend in the number of studies. It takes a greater rise when it goes from 0 in 2012 to 98 publications in 2021, and marks a constant growth trend until said period, which later falls to 77 publications for the year 2022, which is understandable, due to the fact that high-quality journals and publishers have a natural delay in their publication process, so the total number of articles published up to this point may not be reflected in the current year.

Figure 1. Annual scientific production
Source: own creation

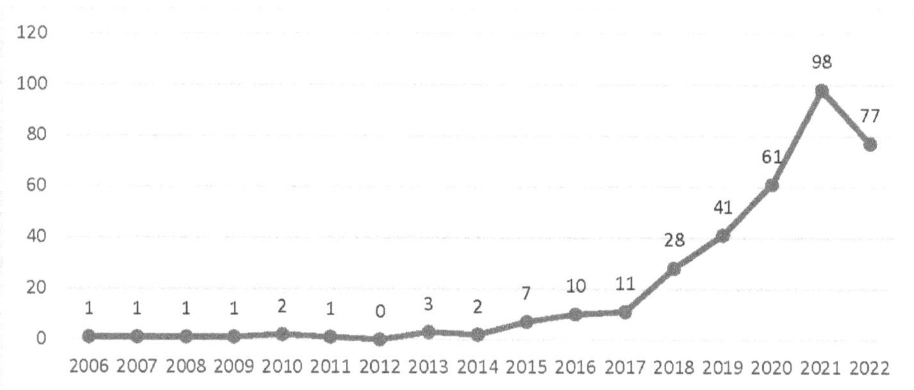

Most Relevant Magazines

Regarding the number of publications on the topic of organizational cybersecurity, Figure 2 shows three journals that concentrate the largest number of publications:

Journal of Cybersecurity with 15, Applied Science-Basel with 12, and Computers & Security Magazine with 11 respectively.

These findings represent an interesting coverage regarding the investigations that have been approached from this perspective within the scientific databases. They show evidence of interest within the scientific community in carrying out studies focused on the aspect of cybersecurity in organizations.

Figure 2. Most relevant sources
Source: own creation

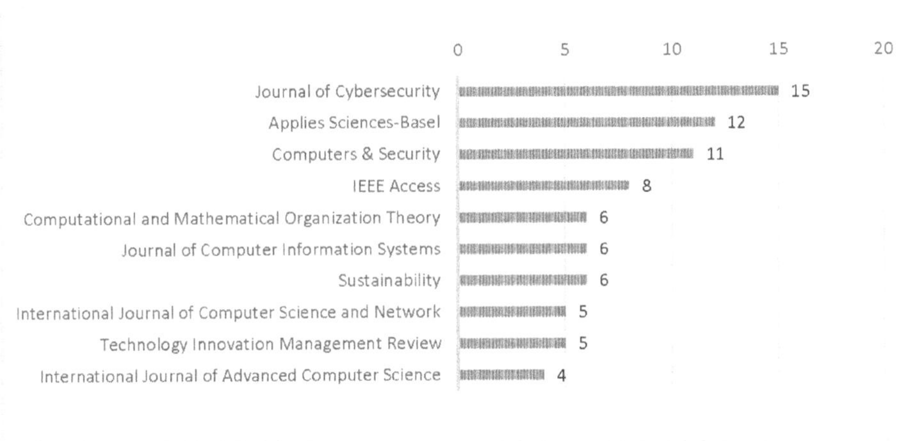

LOCAL IMPACT OF THE SOURCE

In addition to the relevance of the journals with the largest publications on the topic under study, among the bibliometric indicators that allow having a clear picture regarding the scope, as well as their contribution to scientific databases, is the impact they have. These journals have a series of indicators such as the *H-Index*, used to quantify scientific research as well as measure the impact of their citations. The *G-Index*, is introduced as a better element to the *H-Index*, to measure the overall index of citations on a certain set of articles. Finally, the *M-Index*, which is used to facilitate comparisons between academic elements with different scopes, considering the number of years since the first publication (Egghe, 2006). In this sense, Table 1 shows the *top 10* of the journals with the most dominance within the platform, ordered from highest to lowest impact according to the *G-Index* indicator, the total citations of the journal, the number of publications, as well as the year that he began to publish on the subject.

Table 1. Magazine impact

N°	Element	h_index	g_index	m_index	TC	NP	YP_ini
1	Journal of Cybersecurity	6	10	0.85	113	10	2016
2	Applied Sciences-Basel	3	8	0.6	71	9	2018
3	Computers & Security	6	8	1.5	117	8	2019
4	IEEE Access	3	5	0.75	83	5	2019
5	Computational and Mathematical Organization Theory	3	4	1	30	4	2020
6	Sensors	3	4	0.75	49	4	2019
7	Technology Innovation Management Review	3	4	0.3	94	4	2013
8	Applied Network Science	2	3	0.5	19	3	2019
9	Cybersecurity	2	3	0.4	14	3	2018
10	Information and Computer Security	2	3	0.25	111	3	2015

Source: own creation

TC=Total of citazion, NP=Number of publications, YP_ini=Year of publication (initial)

Authors With the Highest Number of Publications

For this analysis, Figure 3 indicates that the distribution of publications is as follows: Carley K. has 9 publications, followed by Cam H., Ganesan S. and Jajodia S. with 5, Shah A. with 4, and finally Belluta D., Convertino M., Crigen D., Jeyaraj A. and Ng L. with 3 each.

Most Relevant Affiliations

Table 2 shows the 10 institutions with the highest number of publications related to organizational cybersecurity. This list is led by Carnegie Mellon University with 24 publications, followed by Deakin University as well as the University of Central Florida with 13 publications each, Hokkaido University with 12, and Warsaw University of Technology with 11 contributions.

An interesting element regarding this bibliometric indicator is that it allows observing the origin of the institutions where contributions are made on the topic, it is clearly observed that the United States of America is the leading country with the largest number of publications within its institutions, as well as there are countries such as Australia, Japan, Poland, Saudi Arabia, South Korea, Russia, Ukraine and Canada, which have institutions that carry out research in the field of cybersecurity.

Figure 3. Authors with the highest number of publications
Source: own creation

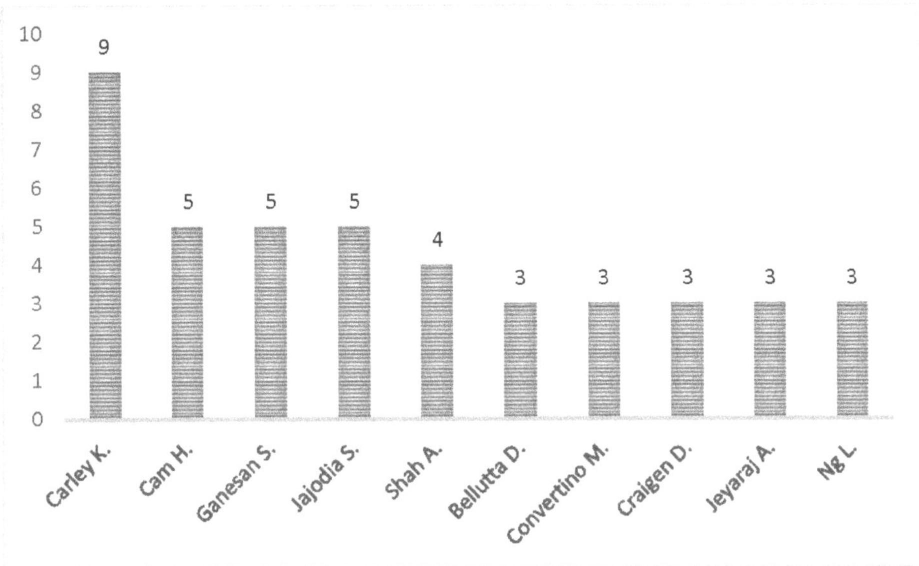

Table 2. Most relevant affiliations

University	Country	Articles
Carnegie Mellon University	United States of America	24
Deakin University	Australia	13
University of Central Florida	United States of America	13
Hokkaido University	Japan	12
Warsaw University Technology	Poland	11
King Saud University	Saudi Arabia	9
Korea University	South Korea	9
University of California San Diego	United States of America	8
The University of Melbourne	Australia	8
The University of South Florida	United States of America	8
Kazan Innovative University	Russia	7
Lviv Polytechnic National University	Ukraine	7
University of Toledo	United States of America	7
Wright State University	United States of America	6
Carleton University	Canada	6

Source: Own Creation

Scientific Production by Countries

With the purpose of complementing Table 2, Table 3 shows the academic production of the 10 countries that are referents on the study of organizational cybersecurity within the Web of Science platform. The list is led by the United States of America with 101 contributions, followed by the United Kingdom with 21, China with 16, Australia with 15, Saudi Arabia with 13, South Korea with 12, Italy with 11, as well as the Netherlands, Russia and Ukraine with 10. The foregoing shows a concentration in a few countries that can set the tone for reducing cybercrime and achieving competitive advantage in the short term.

Table 3. Scientific production by countries

Country	Frequency
United States of America	101
United Kingdom	21
China	16
Australia	15
Saudi Arabia	13
South Korea	12
Italia	11
Netherlands	10
Russia	10
Ukraine	10

Source: Own Creation

Most Cited Documents

Table 4 shows the documents with the highest number of citations, the most relevant article being the publication by Meneghello et al. (2019), which analyzes the Internet of Things (IoT) and how it is expanding within the different areas that surround us, as well as industry 4.0. The above scenario demonstrates the critical role of cybersecurity in the face of security threats related to attack mechanisms, highlighting the importance of cybersecurity as an integral part in the design of information systems.

Subsequently, there is a study by Sun et al. (2018), in which they highlight the increase in the number of incidents related to cyber security, as well as the interest of governments, companies and society in preparing to combat cyber threats. In this sense, they analyze the representative state of the art on cybersecurity incidents, essentially through the use of organizational reports to publish information related to security.

These documents, to mention a few, highlight the importance of the subject, as well as the approach given to it from two different perspectives, focused on its analysis.

Table 4. Most cited documents

N°	Main Author	Article	Magazine	TC	TC per Year	Year
1	Meneghello F.	IoT: Internet of Threats? A survey of practical security vulnerabilities in real IoT devices	IEEE Internet of Things Journal	210	52.50	2019
2	Sun N.	Data-Driven cybersecurity incident prediction: A survey	IEEE Communications Surveys & Tutorials	118	29.25	2018
3	Ullah F.	Cyber security threats detection in Internet of Things using deep learning approach	IEEE Access	70	17.50	2019
4	Leng J.	Blockchain-secured smart manufacturing in industry 4.0: A survey	IEEE Transactions on Systems, Man, and Cybernetics: Systems	68	34.00	2020
5	Homoliak I.	Insight into insiders and IT: A survey of insider threath taxonomies, analysis, modeling, and countermeasures	ACM Computing Surveys	68	17.00	2020
6	Craigen D.	Defining cybersecurity	Technology Innovation Management Review	61	6..77	2014
7	Rees L.	Decision support for cybersecurity risk planning	Decision Support Systems	52	4.33	2011
8	Moore T.	The economics of cybersecurity: Principles and policy options	International Journal of Critical Infraestructure Protection	49	3.76	2010
9	Boyson S.	Cyber supply chain risk management: Revolutionzing the strategic control critical IT systems	Technovation	45	5.00	2014
10	Yang L.	Effective repair strategy against advanced persistent threat: A differential game approach	IEEE Transactions on Information Forensics and Security	45	11.25	2018

continues on following page

Table 4. Continued

N°	Main Author	Article	Magazine	TC	TC per Year	Year
11	Sittig D.	A socio-technical approach to preventing, mitigating, and recovering from ransomware attacks	Applied Clinical Informatics	44	6.28	2016
12	Kure H.	An integrated cyber security risk management approach for a cyber-physical system	Applied Sciences	43	8.60	2018
13	Patel S.	Quantitatively assessing the vulnerability of critical information systems: A new method for evaluating security enhancements	International Journal of Information Management	42	2.80	2008
14	Dawson J.	The future cybersecurity workforce: Going beyond technical skills for successful cyber performance	Frontiers in Psychology	37	7.40	2018
15	Mueller M.	Against sovereignty in cyberspace	International Studies Review	36	12.00	2020

Source: Own creation
TC: Total of Citations

Theories Used

Regarding the analysis of the theories used in the studies, it was kind enough to analyze each of the contents of the articles that were empirical studies and that used theories related to the study of the topic of interest. After this exhaustive review of the documents, the publications that had the requirements for analysis were a total of 17 of the 345 publications found in the database.

Table 5 shows a summary of these articles, as well as each of the theories used in the study of cybersecurity. It is observed that the Theory of Motivation to Protection is the most used (25%), followed by the Institutional Theory (17%) and the Risk Perception Theory (13%), respectively.

Regarding the analysis techniques used, 50% of the studies use Structural Equation Modeling, 31% Linear Regression; as well as the Multiple Criteria Decision Analysis, the Probit Regression Model, the Logistic Regression and the Hierarchical Regression Model with 6% respectively. Regarding the regions where these studies were carried out, they are: The United States of America is where the most studies have been carried out on the subject, followed by studies globally, as well as the United Kingdom, Pakistan, Oman, Jordan, Indonesia, Bahrain, Nigeria, and Slovenia respectively.

Table 5. Summary of articles

Region	Author	Variables	Analysis Unit	Analysis Technique	Theory
Global	Jeyaraj & Zadeh, (2020)	Mimetic pressures, normative pressures and coercive pressures	Annual Reports	Lineal regression	Institutional Theory
USA	Smith, et al., (2021)	Human capital, relational capital, structural capital	Enterprises S&P 500	Lineal regression	Upper Echelons Theory
USA	Philip, et al., (2023)	Cognitive and psychological factors	Employees	Structural equations	Rational Choice Theory Theory of Planned Behavior
United Kingdom	Sukumar et al., (2023)	Security, dependency, strategy, legal aspect	Entrepreneurs	Multiple Criteria Decision Analysis	Risk Perception Theory
USA	Héroux & Fortin, (2022)	IT board expert, board independence, women directors on the board, age of the board of directors, committee responsible for cybersecurity on the board of directors	Enterprises S&P/TSX	Multiple Regression	Upper Echelons Theory
Pakistan	Firdaus et al., (2022)	Perceived pressures, perceived opportunity, rationality, capabilities, exposure	Banking Sector	Multiple Regression	Fraud Triangle Theory
Global	Michalakopoulou et al., (2022)	Human capital, social capital, technology transfer	Legal Sector	Probit Regression Model	Resources and Capacities Theory
Oman	Al-Shanfari et al., (2022)	Attitudes, subjective norms, perceived behavioral control, intention, efficacy, vulnerability, certainty of sanctions, severity of sanctions, organizational support, security of communication channels, security, education, and training.	Public Sector	Structural equations	Protection Motivation Theory, Theory of Planned Behavior, Deterrence Theory
USA	Caldarulo et al., (2022)	Technological Capabilities	SMEs	Logistic Regression	Routine Activity Theory
USA	Anwar et al., (2017)	Perceived vulnerability, perceived severity, trust, perceived barriers, perceived benefits, efficacy, and peer behavior.	Employees	Multiple Regression	Protection Motivation Theory
Global	Blythe & Coventry (2018)	Perceived severity, perceived vulnerability, response costs, experience, organizational citizenship, security accountability.	Employees	Hierarchical Regression Model	Protection Motivation Theory

continues on following page

Table 4. Continued

Region	Author	Variables	Analysis Unit	Analysis Technique	Theory
USA	Li et al., (2019)	Experience, perceived vulnerability, perceived threat, perceived barriers, effectiveness	Employees	Structural equations	Protection Motivation Theory
Jordan	Atwah (2022)	Technical aspects, senior management, coercive pressures, and mimetic pressures	Government	Structural equations	Institutional Theory
Indonesia	Berlilana et al., (2021)	Technological context, organizational context, environmental context: optimism, innovation, insecurity, tangible benefit, intangible benefit.	Professionals IT	Structural equations	Institutional Theory
Bahrain	Hasan, et al., (2021)	Technological infrastructure, management support, organizational skills, organizational culture, Government support, industry standard, regulations.	Professionals IT	Structural equations	Institutional Theory, Game Theory, and Deterrence Theory.
Nigeria	Onumo, (2021)	Leadership, knowledge, attitudes, competitive trait, subjectivity	Professionals IT	Structural equations	Theory of Planned Behavior
Slovenia	Vrhovec & Mihelic (2021)	Perceived severity, perceived vulnerability, perceived threats, self-efficacy, fear of cyber-attacks	Academics	Structural equations	Protection Motivation Theory

Source: Own creation

Collaboration Network by Country

Meanwhile, the collaboration indicator by country, Figure 4, shows its mapping of the collaboration network. In a greater degree of centrality, the United States of America, the United Kingdom, China, Canada and Singapore are located, which, in turn, maintain a collaboration link with the different countries located in the peripheral area of the figure shown.

These collaborations allow us to inquire about the interest of studies related to cybersecurity in organizations, since, for the most part, they are developed countries, in which technology from both governments and private companies play an essential role for the execution of their functions, therefore, since technology is a fundamental element for most institutions, they become vulnerable to receiving cyber-attacks that threaten their privacy. These types of problems have led to the development of research around the search for solutions that help prevent, mitigate and avoid this type of situation, so that institutions, universities, as well as government organizations and research centers, can encourage the development of intervention strategies and proposals.

The bibliometric indicators of the Web of Science platform provide the guideline to observe those agencies and departments that have funding bags for the development of research related to the topic of cybersecurity. Among them are the Office of Naval Research, the National Science Foundation Program, the European Commission, the Center for Informed Democracy & Social-cybersecurity, as well as UK Research and Innovation, in which, according to the *Web of Science*, it is inferred that these organizations had interference in the development of some studies related to the subject.

Figure 4. Collaboration network by country
Source: own creation

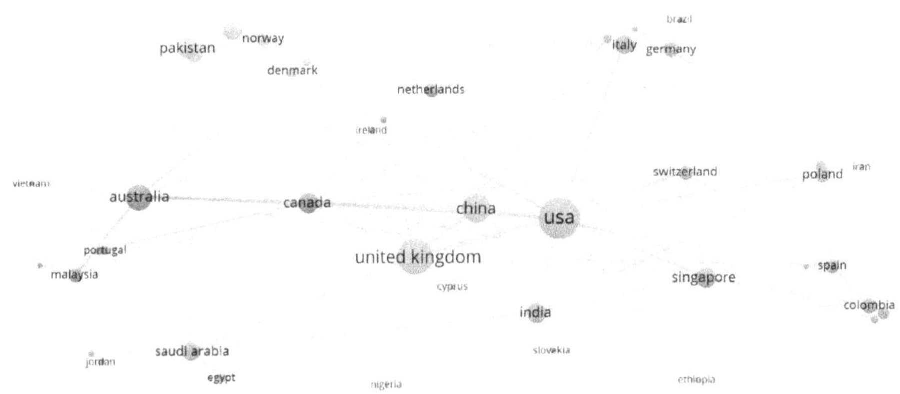

Most Relevant Key Terms

One of the elements that make up the publications within the field of scientific research, are the keywords, their purpose is to delimit the content of the research so that it can facilitate the search for information related to a specific topic within the databases scientific data. Figure 5 shows information related to the key terms most used by researchers in their publications, the term *management* is the one that leads with a total of 29, *security* with 27, these two are followed by the term *information security* with 19, and *cybersecurity* with 18, this ranking is complemented by the term *Internet, systems, impact, challenges, risks, organization, and performance.*

Figure 5. Key terms most used
Source: own creation

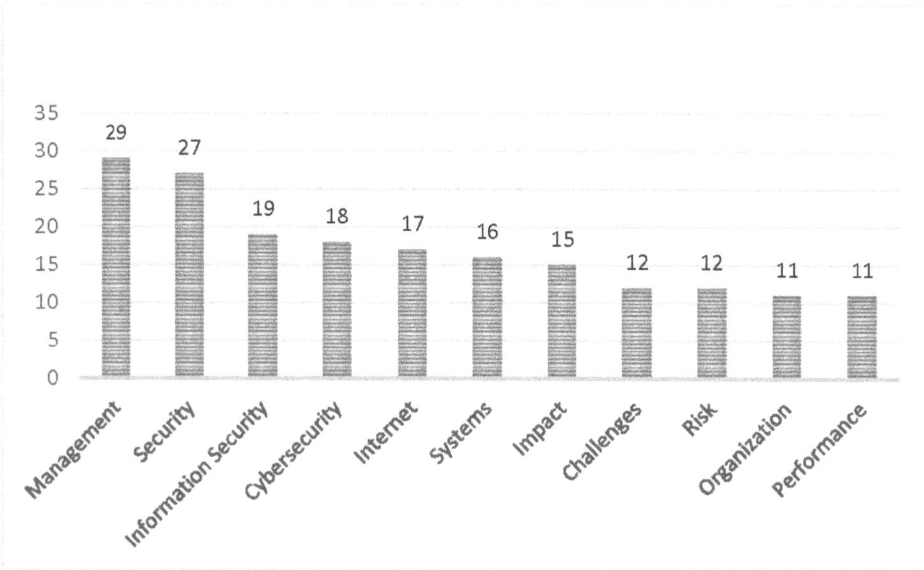

Key Terms Network

Once the information of the most used terminology was extracted, the relationship mapping of all the key terms was carried out with the support of the VOSviewer software (Van Eck & Waltman, 2010), to carry out the grouping and achieve the visualization of the results. The distance between the key terms, as well as the thickness of the link that unites them, reflects the strength of their link, as well as the relationship they have, so that the shorter the distance between the elements, the stronger a certain relationship is.

As previously mentioned, the terms *management, cybersecurity, model, Internet, risks, privacy, information security and performance*, their grouping of elements within the mapping of key terms, go hand in hand with the conceptualization used to obtain the information and carrying out the corresponding analyses. Without a doubt, the adequate administration of the organization's systems through cybersecurity models will allow the prevention of risks related to the violation of privacy and information security, thus reducing the performance of the organization.

Meanwhile, to the clusters of key terms shown in Figure 6, in the first block are the concepts of *architecture, algorithms, attack detection, Internet, IoT, and networks.* The second of the groupings that occurs, starting with *cybersecurity, acceptance, awareness, behavior, deterrence and firm performance.*

The third of them revolves around information security, which is more closely linked to terms such as *adoption, business, privacy, and protection*. The fourth block is related to aspects of the organization and its link with concepts such as *big data, governance, performance, technology and trust.* Lastly, and at a lower level, is the *challenges, design, risk and threats link.*

Figure 6. Network of key terms
Source: own creation

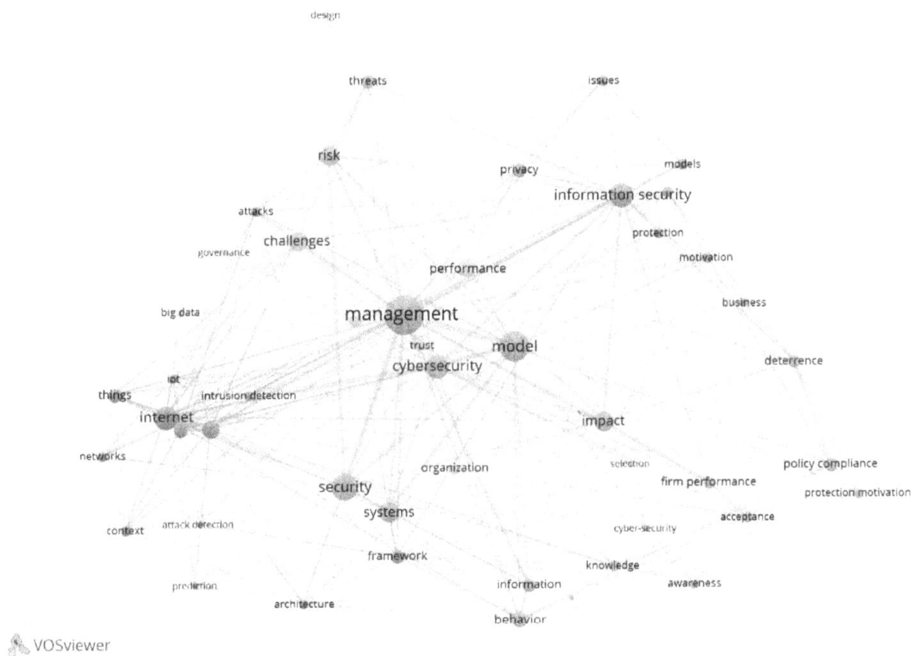

CONCLUSION

The use and implementation of information technologies within organizations entails a series of managerial challenges derived from incidents related to cybersecurity threats within them. This research aims to develop a bibliometric study on the role of cybersecurity from the organizational level within high-impact scientific bases, for which a series of analyzes of bibliometric indicators was developed, in order to infer about what has been investigated on the subject from different areas within the time horizon from which the publications date. Clearly, within the annual scientific production, a growing and constant trend is observed, which allows us to understand the importance of this subject of study within the field of scientific research, especially in recent years.

In addition to the analysis of the publications, the impact of the journals within the platform was analyzed with the *G-Index* indicator, to determine those with a higher degree of quality and influence within the platform, thus representing an area of interest. Interesting opportunity to broaden the spectrum of study to different platforms with related publications from their databases and to be able to compare said impact.

Another consideration is in relation to the analysis of the most important affiliations, which are mostly in developed countries, since there are institutions, organizations and research centers that finance the development of research related to cybersecurity, therefore, this aspect also influences the number of publications by country, as well as the collaboration networks between them.

Among the indicators examined, the most relevant documents are observed by number of citations, in which the first two documents were analyzed, as well as the perspectives from which they were addressed, they are important resources for the study and understanding of the subject and together with the rest of the documents are an interesting base of theoretical support that allows understanding the study phenomenon, as well as the different approaches and contexts where they have been developed and implemented. However, it is necessary to clarify that the analysis of the citations of these documents only covers the articles published within the Web of Science, so it would be interesting to broaden the spectrum of study to other databases and observe the behavior of these metrics in specific.

Regarding the theories used in the study of the subject, they demonstrate the variety of approaches and elements available to analyze and address the challenges related to the study of cybersecurity within organizations. These theories offer a wide range of perspectives and tools to understand the behavior of the actors involved, the motivations behind criminal actions in computer systems, as well as strategies to prevent and mitigate cyber risks. By incorporating these theoretical approaches into the study of cybersecurity, understanding of the underlying factors in the cybersecurity environment can be improved, allowing more effective policies and practices to be designed to protect organizations and individuals from growing threats in the cybersecurity environment line. The use of these theories in empirical studies within the scientific community, highlight the importance of a holistic and multidisciplinary vision in the research and practice of cybersecurity in organizations.

Finally, within the mapping of the network of key terms, the importance is highlighted as well as the link between each of the elements for carrying out the corresponding analyses. Without a doubt, the adequate administration of organizational information systems, through the implementation and use of cybersecurity models, will allow risk prevention, minimizing the vulnerability of both privacy and information security, which can undermine the performance of the organization. Therefore, the use of adequate algorithms within the software architecture, can support the detection of cyber-attacks and prevent future risks, and with this, it is necessary, both the acceptance and adoption of cybersecurity, as well as being aware regarding the benefits that its implementation entails, with the purpose of deterring attacks and cyber threats that could impact the performance of the organization.

Undoubtedly, cybersecurity is a topic that is booming, especially due to the dependence on technology to carry out activities in most sectors of society. No institution is exempt from suffering threats or attacks that violate information security, and as was seen in the development of the investigation, it is a latent issue, especially for the characterization of the phenomenon from different angles that allows its holistic understanding for the organizations, and continue with the development of research that contributes to the field of study.

Limitations: The study has a series of interesting opportunities to develop as future lines of research. Only journals indexed in the *Journal Citations Reports* of the *Web of Science* were considered; publications from different scientific platforms that host related studies such as *Scopus, JStor, EbscoHost*, to name a few, are not included, as well as the integration of indicators other than those here considered.

An area of opportunity is related to the scientific mapping of information, its use in the present study complements the findings made, due to the fact that they are dynamic analyses, and are subject to change over time due to the enrichment of the state of the art derived from new studies and contributions to knowledge.

REFERENCES

Ahmad, A., Maynard, S. B., Desouza, K. C., Kotsias, J., Whitty, M. T., & Baskerville, R. L. (2021). How can organizations develop situation awareness for incident response: A case study of management practice. *Computers & Security*, *101*, 102122. doi:10.1016/j.cose.2020.102122

Al-Qahtani, A. F., & Cresci, S. (2022). The COVID-19 scamdemic: A survey of pishing attacks and their countermeasures during COVID-19. *IET Information Security*, *16*(5), 324–345. doi:10.1049/ise2.12073 PMID:35942004

Al-Shanfari, I., Yassin, W., Tabook, N., Ismail, R., & Ismail, A. (2022). Determinants of Information Security Awareness and Behaviour Strategies in Public Sector Organizations among Employees. *International Journal of Advanced Computer Science and Applications*, *13*(8). doi:10.14569/IJACSA.2022.0130855

Anwar, M., He, W., Ash, I., Yuan, X., Li, L., & Xu, L. (2017). *Gender difference and employees' cybersecurity behaviors*. Comptuters in Human Behavior., doi:10.1016/j.chb.2016.12.040

Aria, M., & Cuccurullo, C. (2017). Bibliometrix: An R-tool for comprehensive science mapping analysis. *Journal of Informetrics, 11*(4), 959–975. doi:10.1016/j.joi.2017.08.007

Atwah, A. M. (2022). Investigating the drivers of cybersecurity enhancement in public organizations: The case of Jordan. *The Electronic Journal on Information Systems in Developing Countries*, *88*(5), 12223. doi:10.1002/isd2.12223

Berlilana, N., Noparumpa, T., Ruangkanjanases, A., Hariguna, T., & Sarmini. (2021). Organization benefit as an outcome of organizational security adoption: The role of cyber security readiness and technology readiness. *Sustainability (Basel)*, *13*(24), 13761. doi:10.3390u132413761

Blythe, J. M., & Coventry, L. (2018). Costly but effective: Comparing the factors that influence employee anti-malware behaviours. *Computers in Human Behavior*, *87*, 87–97. doi:10.1016/j.chb.2018.05.023

Boyson, S. (2014). Cyber supply chain risk management: Revolutionizing the strategic control of critical IT systems. *Technovation*, *34*(7), 342–353. doi:10.1016/j.technovation.2014.02.001

Caldarulo, M., Welch, E. W., & Feeney, M. K. (2022). Determinants of cyber-incidents among small and medium US cities. *Government Information Quarterly*, *39*(3), 101703. doi:10.1016/j.giq.2022.101703

Carrapico, H., & Farrand, B. (2020). Discursive continuity and change in the time of Covid-19: The case of EU cybersecurity policy. *Journal of European Integration*, *42*(8), 1111–1126. doi:10.1080/07036337.2020.1853122

Craigen, D., Diakun-Thilbault, N., & Purse, R. (2014). Defining cybersecurity. *Technology Innovation Management Review*, *4*(10), 13–21. doi:10.22215/timreview/835

Dawson, J., & Thomson, R. (2018). The future cybersecurity workforce: Going beyond technical skills for successful cyber performance. *Frontiers in Psychology*, *9*, 744. doi:10.3389/fpsyg.2018.00744 PMID:29946276

Dzyana, H., Pasichnyk, V., Garmash, Y., Naumko, M., & Didych, O. (2022). The System for Ensuring the Information Security of the Organization in the Context of Covid-19 Based on Public-Private Partnership. *International Journal of Computer Science and Network Security*, 19-24. doi:10.22937/IJCSNS.2022.22.6.4

Egghe, L. (2006). Theory and practise of the g-index. *Scientometrics, 69*(1), 131–152. doi:10.100711192-006-0144-7

FBI. (2021). *Internet Crime Complaint Center IC3.* FBI. https://www.ic3.gov/

Firdaus, R., Xue, Y., Gang, L., & Ali, M. S. (2022). Artificial Intelligence and Human Psychology in Online Transaction Fraud. *Frontiers in Psychology, 13*, 947234. Advance online publication. doi:10.3389/fpsyg.2022.947234 PMID:36304851

Furnell, S., Haney, J., & Theofanos, M. (2021). Pandemic Parallels: What can cybersecurity learn from Covid-19? *Computer, 54*(3), 68–72. doi:10.1109/MC.2020.3046888 PMID:34385715

Goyanes, M., & Demeter, M. (2020). How the geographic diversity of editorial boards affects what is published in JCR-Ranked communication journals. *Journalism & Mass Communication Quarterly, 97*(4), 1–26. doi:10.1177/1077699020904169

Hallinger, P., & Kovačević, J. (2019). Science mapping the knowledge base in educational leadership and management: A longitudinal bibliometric analysis, 1960 to 2018. *Educational Management Administration & Leadership, 49*(1), 5–30. doi:10.1177/1741143219859002

Hasan, S., Ali, M., Kurnia, S., & Thurasamy, R. (2021). Evaluating the cyber security readiness of organizations and its influence on performance. *Journal of Information Security and Applications, 102726*, 102726. doi:10.1016/j.jisa.2020.102726

Hernandez, J. G., & Salcedo, M. T. (2020). The influence and benefits of technology as a strategy in organizations. *Multidisciplinary Journal for Education Social and Technological Sciences, 7*(1), 32–53. doi:10.4995/muse.2020.10693

Héroux, S., & Fortin, A. (2022). Board of directors' attributes and aspects of cybersecurity disclosure. *The Journal of Management and Governance*. doi:10.100710997-022-09660-7

Homoliak, I., Toffalini, F., Guarnizo, J., Elovici, Y., & Ochoa, M. (2020). Insight into insiders and IT: A Survey of insider threat taxonomies, analysis, modeling, and countermeasures. *ACM Computing Surveys, 52*(30), 1–40. doi:10.1145/3303771

Hu, W., Chang, C., Sengupta, A., Bhunia, S., Kastner, R., & Li, H. (2021). An overview of hardware security and trust: Threats, countermeasures, and design tools. *IEEE Transactions on Computer-Aided Design of Integrated Circuits and Systems*, *40*(6), 1010–1038. doi:10.1109/TCAD.2020.3047976

Jafar, M. T., Al-Fawa'reh, M., Barhoush, M., & Alshira'H, M. H. (2022). Enhanced Analysis Approach to Detect Pishing Atacks During COVID-19 Crisis. *Cybernetics and Information Technologies*, *22*(1), 60–76. doi:10.2478/cait-2022-0004

Jeyaraj, A., & Zadeh, A. (2020). Institutional isomorphism in organizational cybersecurity. *Journal of Organizational Computing and Electronic Commerce*, *20*(4), 361–380. doi:10.1080/10919392.2020.1776033

Kim, N., & Lee, S. (2021). Cybersecurity breach and crisis response: An analysis of organizations' official statements in the United States and South Korea. *International Journal of Business Communication*, *58*(4), 560–581. doi:10.1177/2329488418777037

Kipper, L. M., Furstenau, L. B., Hoppe, D., Frozza, R., & Lespen, S. (2019). Scopus scientific mapping production in industry 4.0 (2011-2018): A bibliometric analysis. *International Journal of Production Research*, 1–24. doi:10.1080/00207543.2019.1671625

Kure, H. I., Islam, S., & Razzaque, M. A. (2018). An Integrated Cyber Security Risk Management Approach for a Cyber-Physical System. *Applied Sciences (Basel, Switzerland)*, *8*(6), 898. doi:10.3390/app8060898

Latilla, V. M., Urbinati, A., Cavallo, A., Franzo, S., & Ghezzi, A. (2021). Organizational re-design for business model innovation while exploiting digital technologies: A single case study of an energy company. *International Journal on Management of Innovation & Technology*, *18*(2), 2040002. doi:10.1142/S0219877020400027

Leng, J., Ye, S., Zhou, M., Zhao, J. L., Liu, Q., Guo, W., Cao, W., & Fu, L. (2020). Blockchain-Secured Smart Manufacturing in Industry 4.0: A Survey. *IEEE Transactions on Systems, Man, and Cybernetics. Systems*, *51*(1), 237–252. doi:10.1109/TSMC.2020.3040789

Li, L., He, W., Xu, L., Ash, I., Anwar, M., & Yuan, X. (2019). Investigating the impact of cybersecurity policy awareness on employees' cybersecurity behavior. *International Journal of Information Management*, *45*, 13–24. doi:10.1016/j.ijinfomgt.2018.10.017

Loan, F. A., Bisma, B., & Nahida, N. (2022). Global research productivity in cybersecurity: A scientometric study. *Global Knowledge. Memory and Communication*, *7*(4/5), 342–354. doi:10.1108/GKMC-09-2020-0148

Meneghello, F., Calore, M., Zucchetto, D., Polese, M., & Zanella, A. (2019). IoT: Internet of Threats? A Survey of Practical Security Vulnerabilities in Real IoT Devices. *IEEE Internet of Things Journal*, *6*(5), 8182–8201. doi:10.1109/JIOT.2019.2935189

Michalakopoulou, K., Nikita, A., Njoya, E. T., & Johnes, J. (2022). Innovation in the legal service industry: Examining the roles of human and social capital, and knowledge and technology transfer. *International Journal of Entrepreneurship and Innovation*. doi:10.1177/14657503221119667

Moore, T. (2010). The economics of cybersecurity: Principles and policy options. *International Journal of Critical Infrastructure Protection*, *3*(3-4), 103–117. doi:10.1016/j.ijcip.2010.10.002

Morgan, S. (2020). Cybercrime to cost the world $10.5 trillion annually by 2025. *Cybercrime Magazine*. https://cybersecurityventures.com/hackerpocalypse-cybercrime-report-2016/

Mueller, M. L. (2020). Against sovereignty in cyberspace. *International Studies Review*, *22*(4), 779–801. doi:10.1093/isr/viz044

Ng, I. C., & Wakenshaw, S. Y. (2017). The Internet-of-Things: Review and research directions. *International Journal of Research in Marketing*, *34*(1), 3–21. doi:10.1016/j.ijresmar.2016.11.003

Onumo, A., Ullah-Awan, I., & Cullen, A. (2021). Assessing the moderating effect of security technologies on employees compliance with cybersecurity control procedures. [TMIS]. *ACM Transactions on Management Information Systems*, *12*(2), 1–29. doi:10.1145/3424282

Park, Y., El Sawy, O. A., & Fiss, P. C. (2017). The role of business intelligence and communication technologies in organizational agility: A configurational approach. *Journal of the Association for Information Systems*, *18*(9), 648–686. doi:10.17705/1jais.00467

Patel, S., Graham, J., & Ralston, P. (2008). Quantitatively assessing the vulnerability of critical information systems: A new method for evaluating security enhancements. *International Journal of Information Management*, *28*(6), 483–491. doi:10.1016/j.ijinfomgt.2008.01.009

Philip, S. J., Luu, T., & Carte, T. (2023). There's no place like home: Understanding users' intentions toward securing Internet-of-thing (IoT) smart home networks. *Computers in Human Behavior*, *107551*(139), 107551. doi:10.1016/j.chb.2022.107551

Pranggono, B., & Arabo, A. (2020). COVID-19 Pandemic cybersecurity issues. *Internet Technology Letters*, *4*(2). doi:10.1002/itl2.247

Rees, L. P., Deame, J. K., Rakes, T. R., & Baker, W. H. (2011). Decision support for cybersecurity risk planning. *Decision Support Systems*, *51*(3), 493–505. doi:10.1016/j.dss.2011.02.013

Sittig, D. F., & Singh, H. (2016). A socio-technical approach to preventing, mitigating, and recovering from ransomware attacks. *Applied Clinical Informatics*, *7*(2), 624–632. doi:10.4338/ACI-2016-04-SOA-0064 PMID:27437066

Smith, T., Tadesse, A., & Vincent, N. E. (2021). The impact of CIO characteristics on data breaches. *International Journal of Accounting Information Systems*, *100532*, 100532. doi:10.1016/j.accinf.2021.100532

Snyder, H. (2019). Literature review as a research methodology: An overview and guidelines. *Journal of Business Research*, *104*, 333–339. doi:10.1016/j.jbusres.2019.07.039

Stetsenko, S., Tytok, V., Emelianova, O., Bielienkova, O., & Tsyfra, T. (2021). The Interrelation of Digital Technologies and Organizational and Economic Mechanisms in Construction: Adaptation to Change Management. *International Review (Steubenville, Ohio)*, *1-2*(1), 21–31.

Sukumar, A., Mahdiraji, H. A., & Jarari-Sadeghi, V. (2023). Cyber risk assessment in small and medium-sized enterprises: A multi-level decision-making approach for small e-tailors. *Risk Analysis*, risa.14092. doi:10.1111/risa.14092 PMID:36627823

Sun, N., Zhang, J., Rimba, P., Gao, S., Zhang, L. Y., & Xiang, Y. (2018). Data-Driven Cybersecurity Incident Prediction: A Survey. *IEEE Communications Surveys and Tutorials*, *21*(2), 1744–1772. doi:10.1109/COMST.2018.2885561

Ullah, F., Naeem, H., Jabbar, S., Khalid, S., Latif, M., Al-Turjam, F., & Mostarda, L. (2019). Cyber security threats detection in internet of things using deep learning approach. *IEEE Access : Practical Innovations, Open Solutions*, *7*, 124379–124389. doi:10.1109/ACCESS.2019.2937347

Van Eck, N. J., & Waltman, L. (2010). Software survey: VOSviewer, a computer program for bibliometric mapping. *Scientometrics*, *84*(2), 523–538. doi:10.100711192-009-0146-3 PMID:20585380

Vrhovec, S., & Mihelic, A. (2021). Redefining threat appraisals of organizational insiders and exploring the moderating role of fear in cyberattack protection motivation. *Computers & Security*, *106*, 102309. doi:10.1016/j.cose.2021.102309

Yang, L., Li, P., Zhang, Y., Yang, X., Xiang, Y., & Zhou, W. (2018). Effective Repair Strategy Against Advanced Persistent Threat: A Differential Game Approach. *IEEE Transactions on Information Forensics and Security*, *14*(7), 1713–1728. doi:10.1109/TIFS.2018.2885251

Yong, L. A., Rivas, L. A., & Chaparro, J. (2010). Technological acceptance model (TAM): A study of the influence of the national culture and of the user profile in the use of ICTS. *Innovar (Universidad Nacional de Colombia)*, *20*(36), 187–203.

Chapter 9
Factors That Alter Copper Dissolution on Solder SAC 305 From Solder Pots Located on Wave Solder Equipment

Edgar Benjamín García Viramontes
IIES, Mexico

ABSTRACT

An overview on relation between wave solder equipments (WSM) and their users' practices on productive environments that could affect the copper concentration on its solder baths, including external conditions identified on industry environment to create a perspective that differs from scientific studies statements performed on laboratories with controlled conditions. It also aims to establish a link between the new solder bar and recovered solder with the copper dissolution control as the main contributors on a productive environment to keep in control solder alloys on SAC 305.

INTRODUCTION

The industry has been raised up their operations manufacturing different products for certainly purpose, the electronic assemblies has been focusing on development panel circuit boards (PCB) through different process until complete an assigned product which accomplish specs required by customer for usage and comfort on its daily activities.

DOI: 10.4018/978-1-6684-8088-5.ch009

A main useful technology for mass Printed Circuit Boards (PCB) soldering on through hole areas; are the Wave Solder equipments, their usage provides to industry the key to accomplish with daily demands in a lower time compared with manually process costs. Its structure can be defined as follow:

1) Conveyor to move PCB trough the equipment at different zones.
2) Flux application by a Nozzle Fluxer.
3) Pre-Heater Zones assigned to activate flux chemicals and avoid thermal shock when pass through the Wave.
4) Wave Soldering area (Solder POT) in which one occurs the intermetallic process between PCB pads and solder.

The Solder POT design its conformed an structure to support high temperatures surrounded by heating elements with the purpose to apply heat and melt solder bars contained into it and exposing them at temperatures over 260°C (depends from product data sheet and manufacturing requirements); when the solder contained on pot reach a liquid phase its moved by pumping system and nozzles to flow solder as a wave fall which receive the PCB to make contact between them and do able intermetallic process(Solder and PCB pads).

In the past during the first electronic assemblies on industry, the solders used on process contains lead due its properties to facilitate the intermetallic process for solder joints; however, their impact on human health and risks on environment demands replace solders alloy with a Sn, Ag and Cu in different portions that usually solder suppliers identify as SAC (Combination of tin, silver and copper) in different versions as SAC 305,SAC300,SAC 307, as others... Quality, types and welding forms vary according to the models and alloy type makes them up and can be delivered in roll presentations of different thickness diameters, bars with a specific weight, concentration of alloys, according to requirements. Nevertheless, the lead-free solders usage involves changes on process and terms statement for process in order to keep quality.

The term copper dissolution refers to the general phenomenon whereby solid copper leaves most of it when it comes in contact with molten solder, regardless of the cause. Copper erosion strictly refers to the purely mechanical removal of copper, while copper leaching is a chemical-based dissolution. The dissolution of copper is not caused by a single phenomenon, but is the effect of a mixture of causes and variables. (D Di Maio, 2008)

LITERATURE REVIEW

The Copper concentration becomes the biggest challenge to face on manufacturing processes due their impact on the solder alloys affects their properties to join into the PCB pads during the intermetallic process, as the solder alloy contain this metal in assigned percentage its excess turns the solder viscous affecting its flow properties expected as the observed on its liquid phase causing solder bridges or insufficiencies on the process exalting costs added to process to rework this kind of defects.

To keep Solder alloys under control its typically performed a periodical inspection from solder POTs extracting samples to evaluate them by laboratory; to analyze solders and determinate if metals that compound the alloy is under control demands wait studies results to know its response and then apply actions to equilibrate the formula depending the element detected as out of range; during that lapse time the equipment still in use makes to machine users uncertainty at the moment to correct solder failures if they don't have information on time to eliminate factors and identify a root cause .

The behavior from SAC welding contemplates copper concentrations that increase due to the contact interaction between the copper tablet and the welding crucible, periodically evaluating samples that accumulate material exposed to contact times during the IMC generated between the welding tracks. copper from a board and the solder pot at a certain temperature, which increases the requirement to add additives to compensate for the chemical composition (Craig Hamilton, 2007).

Otherwise, it Will be helpful predict a result to let users prevent abnormalities on solder alloys and assurance the best performance. To find a manner and prevent an unbalanced alloy its necessary examinate the process and identify factors involved on it, to detect the biggest contributors that could alter the cupper concentration and determinate which ones can help to balance the compound. Based on productive environment from factory position can be determinate the following elements:

1) WSM Settings: Wave Rpm, Pot height, Wave Turbulence, Pot Temperature, others.
2) Solder Condition: New Solder Bar vs Recycle Solder bar.
3) Environment: Humidity vs Temperature
4) PCBs: Quantities to be processed. Geometry and areas to be soldered, PCB type.

WSM Settings: Wave RPM, Pot Height, Wave Turbulence, Pot Temperature, Others

The main purpose for Wave Solder Machines its prepare and solder through hole components placed into PCB avoiding a thermal shock that may cause delamination or damages, to reach this result its necessary review the profile temperature to be used according to product requirements including the equipment set up to prepare the contact between the solder wave fall with the bottom side of the PCB.

The Pre heater zones assigned on the equipment warm the PCB gradually to prepare its temperature before arrives to solder wave, due to this it's possible reduce high temperature change at the moment to be soldered without damages on product. Aside part of pre heaters function it's to activate the flux applied on the first stage (Nozzle Fluxer) to optimize the solder results.

As Solder POT needs to be adjusted the RPM and POT height are factors involved on the equipment to create a solder wave fall enough liquid to complete intermetallic joint; according to product thickness to be soldered this factors are adjusted and standardize for manufacturing purposes, the turbulence generated on the process applies a friction indirectly on copper pads located on PCB which generate the dissolution from this metal on the solder bath assuming a relation between turbulence of the wave and Copper dissolution on solder bath.

Starting from a welding process, the contact time, welding temperature, welding dynamics and alloys it contains are established as the main agents that condition the dissolution of copper; mentioned agents can be defined in quantifiable units that are prepared with the purpose of simulating scenarios to visualize the effects of the main factors in the process. (Showalter, Sibell, Jean, Klimah, & Biggs, 2016).

The influence of copper concentration through the interaction between variables such as solder flux, temperature and dissolution of the PCB denote as a result a direct relationship between the immersion or contact time of the PCB with the concentration in the solder bath. (Tanabe, 2007)

Solder Condition: New Solder Bar vs. Recycle (Recovered) Solder Bar

Currently exist different lead free solder brands on market which offers alloy compounds variety to fill solder baths from wave solder equipments; for daily productions its required to keep the solder level to avoid manipulate the rpms from motor pump which cause higher turbulence if the cupper dissolution needs to be controlled, but industries put more attention to care the new solder usage consume due costs that implies keep solder level on the pot due the daily consume during the soldering process and dross generation which can be explained as the result for liquid solder interaction with the environment oxygen, for this factor some equipments adds Nitrogen system to cover solder bath surface to isolate the liquid solder from environment and reduce the dross generation; also has been implemented as second source to reduce costs use recovered solder extracted from dross reducing the new solder bar usage on daily process.

The recovered solder it's a key contributor for copper concentration; to explain better its only keep on mind meanwhile the PCBs are soldered on the equipment the intermetallic joint between the solder and PCB pads the copper percentage accumulates on the solder alloy and the extracted recovered solder keeps the same copper concentration accumulated due PCBs processed; so its usage compared to new solder bar needs significant lower, even some users apply a simple rule 80% new and 20% recycle solder to control alloy compound from solder bath, but this not effective in most of the cases, for a real control needs to be defined the impact for recycle solder usage and define strictly the quantities to use daily which demands stablish a system to predict and define new and recovered solder to control and reduce its usage. Various analyzes of the slag reveal that it consists of a mixture of oxides and metallic solder, normally constituting more than 90% of the latter. The presence of the oxide with which the unoxidized metal is initially mixed prevents it from actually meeting with the solder in the bath (Paul & Harris, 2000).

To recover solder contained in the slag, the use of a machine that melts it is applied, and by means of mechanical compression against a perforated wall, the liquid metal is squeezed, separating it from the solid oxide component. However, impurities in the slag are returned to the solder pot, which may result in the need for more frequent bath changes. Because reactive contaminants (eg zinc, aluminum, etc.) tend to segregate preferentially in the slag. On the other hand, more noble materials, such as copper, gold or silver (which dissolves easily in molten solder), tend to remain in the solution and become present in both factions (EVS, 2000).

Environment: Humidity vs. Temperature

Commonly this kind of factors are the last ones observed on process, their control depends from environment and could affect the condition of the solder if their storage it is not correct. However, machine users adjust parameter settings from the equipment to compensate deficiencies observed on the solder process.

As recommendation it will be helpful assign a room with humidity and temperature controlled, including evaluate expiration data to get the best results on the daily performance and inventory control, saving costs avoiding wastes of resources. Also, do not forget review the supplier manual to install a WSM equipment to evaluate the acceptance levels for humidity and temperature from environment that will not affect its performance to get a reference.

At the microstructural level, the hardness in lead-free solders that interact with solders at high temperatures, with a previous degradation or aging process modifying the ambient temperature, promotes a metastable condition, so that the environment (temperature and humidity) contributes to the IMC phase during the process (Ariga, 1999) likewise, solders stored even at room temperature suffer degradation due to the aging effect due to the mechanical properties and microstructure of solder alloys depending on their thermal history, in this sense it is argued that a solder exposed to higher temperatures high and higher cooling speeds generate resistance and hardness at all aging times; likewise, slower cooling encourages a softer and more stable inherent lamellar eutectic weld (Lampe, 1976).

PCBs: Quantities to be Processed, Geometry and Areas to be Soldered, PCB Type

During the soldering process the Intermetallic joint between printed circuits board pads and Solder occurs a metals transference from solder to cupper pads and reverse, causing that cupper concentration from solder bath increases on, due to this the areas exposed and the geometry applied on PCBs to be soldered are factors that influence on the alloy.

Particularly, areas assigned to be soldered support a certainly maximum temperature for certainly time. by its design, which needs to be compatible with components to be soldered to stablish a reference point as limit that could help identify the best parameter set up to prevent delamination or damages on products to be soldered. As the cupper properties has a certainly reaction with environment some PCBs manufacturers apply an special treatment of the copper foil surface on the printed circuit board (PCB), on which a protective film is applied that is easily removed at subsequent soldering temperatures. during contact with the solder wave from the crucible. Molten solder can contain a clean copper surface, immediately to form a solid point quickly, this treatment defined as oxidation foil on the cupper Surface cares PCB pads until be necessary expose them on the soldering process at the moment to make contact with solder wave.

In this way, a linear relationship can be established with the immersion time, type of welding and temperature to which it is exposed for the dissolution effect, which denotes an increase in increasing immersion times with temperature, which makes it easier for the atoms coming from of the metals dissolve in the mixture (Yen et al., 2008).

Explore each Item can be difficult on a productive environment due commitment to accomplish manufacturing targets under quality criteria statements as priority, however if examinate closely the failures record and detect an incensement anomalies in certainly periods without a defined root cause due there's no changes on recipes, damaged parts or any abnormalities involved on that time, probably the solder did not work properly to join into the cupper pads due their alloy lost properties and the equipment users never notice or ignore the problem once magically the symptom disappear. To get a quickly solution can be decided to replace solder bath entire with new solder but the costs involved in this action it's not Beneficent for manufacturers; on other hand if the solder keeps its compounds balanced and controlled in normal levels can be assurance a performance as the solder replacement on Solder Pot but cheaper and easier.

The Copper dissolution is a problem related to lead free solders due its alloy properties on their melting points, which requires higher temperatures than the ones based on Pb including copper concentration, due this factor on excess can cause viscosity causing not be able for intermetallic process. After consider the last facts that could occur on productive environment, directly or indirectly affects the solder alloys contained on solder pot, due to this is necessary contemplate supplier recommendations with the observed performance on process and apply restrictions where could be necessary and identify the factors that represent the biggest contributors to alter the cupper concentration on the alloys to initialize their control.

Its more effective to prevent instead to correct. Actually, exist different statistical tools applied on industry with the purpose to examinate information from process indicators to identify if exist a relation between the variables involved to define their impact on the response and understand which factor could be the major influencer; the linear regression let handle the information with sampling usage which could let predict the response of the process expressed on a equation. Particularly adding an observation during a process focus on Wave solder equipment in which one the main factors identified were the solder new vs Recycle Solder has been the following equation with an impact 93.23% described as follow:

Equation 1. Copper Dissolution Formula.

$$\% \ Cu = 0.5728 - 0.001245 \ New\ Solder\ kg + 0.005189 \ Recycle\ Solder\ kg$$

Figure 1. Lineal prediction from copper dissolution equation

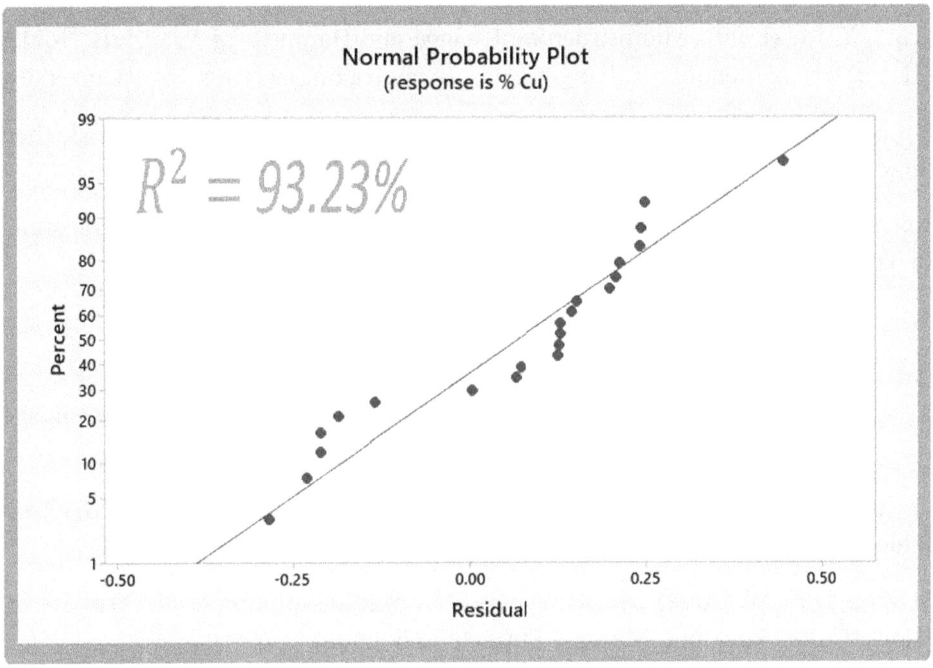

CONCLUSION

After studying and evaluating the factors that intervene in the process, it is concluded that the recycled solder increases the % of Cu in the alloy contained in Solder POT, because it accumulates residuals of this metal from the electronic boards that are being produced by the Inter metallic process, so its use must be measured in a lesser proportion than the new solder, which can reduce the % of Cu accumulated by the processing of electronic boards.

Studies performed on Laboratories are developed on environments controlled, aside the programmed simulations to generate variable interaction between each other to define its potential impact on the system; however, a productive environment involves interactions that differs from laboratories due the production demands, solder consume, recipes and others are in constantly change. Factors established as contributors of scientific researches remark the fundamentals to initialize an analysis to find the biggest contributors on productive environment making possible control them.

REFERENCES

Ariga, Y. M. (1999). Microstructural Change and Hardness of Lead Free Solder Alloys. Dept. of Metallurgical Engineering, School of Engineering, Tokai University.

Willis, D. D. (2008). *Measurements of Copper Disolution in lead free solders*. United kingdom: Depratment for Inovation Universities & Skills.

Bokstein, B. S. A. O. (2011). Gain Boundary Difussion,Stresses and segregation. Switzerland: Trans Tech Publications.

Byle, F. J. (2006). A Study of Copper Dissolution in Pb-Free Solder Fountain Systems. *PanPacific Microelectronics Symposium.*

David, R., & Anderson, D. J. (2012). *Estadistica para negocios y economia*. Cengage Learning.

David Hillman, R. W. (2018). Dissolution Rate of Specific Elements in SAC305 Solder. *Proceedings of SMTA International.*

Di Maio, D. B. W. (2008). Measurements of Copper Dissolution in lead Free Solder Alloys. United kingdom: National Physical Laboratory.

Drápala, P. K. (2012). *Study of Interaction Between*. Salford University.

EVS. (2000). *EVS International Recovery solder*. Obtenido de Leaders on Recovery Solder. http://www.solderrecovery.com/spanish/sq4.html

Goro Izuta, T. T. (2009). Simplified Measuring Method of Copper Concentration in solder bath utilizing copper disolution. *IEEE Transactions on Electronics Packaging Manufacturing*, *32*(3), 138–143. doi:10.1109/TEPM.2009.2021259

Hamilton, C. A. (2007). Have a High copper Dissolution Rates of SAC305/405 Alloys Forced. *SMTAi, PanPacific Microelectronics Symposium.*

Hunt, B. W. (2008). Reduce Cupper Disolution on Lead-Free Assembly. National Physic Laboratory.

Hunt, D. D. (2009). On the factors affecting the dissolution of copper in molten lead free solders and development of a method to asses the soldering parameters. Emerald Group Publishing Limited.

Lampe, B. T. (1976). Room Temperature Aging propierties of some solder alloys. *Welding Journal*, 330s–340s.

Levin, R. I. (2006). Capitulo 7. Muestreo y distribuciones de muestreo. In R. I. Berenso (Ed.), *Estadistica para la administracion (págs* (pp. 206–214). Pearson.

Mageswaran Arunasalam, Z. B. (2021). Challenges in Minimizing Copper Dissolution for Lead Free Wave in surface mount technology going towards green manufacturing. *International Journal of Precision Engineering and Manufacturing-Green Technology.*

Manko, H. I. (1992). *Solders and Soldering*. Mc Graw Hill.

Massalski, T. (1996). *Binary Alloy Phase Diagrams.* Ohio: 2da Edicion.

Satyanarayan, K. P. (2002). *Wetting Behavior of Lead-Free Solders On*. Department of Metallurgical & Materials Engineering National Institute of Technology Karnataka Surathkal,

Showalter, C., Sibell, S., Jean, D., Klimah, P., & Biggs, J. (2016). A Study of Overcoming Solder Icicling And Copper Wire Dissolution. In *An Automated Lead-Free Soldering System*. Kester Inc.

Spajke, R. B. (2013). *Dissolution of a Copper Wire During a Hot Dipping Process Using Sncu1 Lead Free Solder*. Institute of Metals and Technology.

Tanabe, G. I. (2007). *Dissolution of copper on Sn-Ag-Cu system lead*. Emerald Group.

Tetsuro, N. &. (2001). *A control Method for Copper Content in a Solder Dipping Bath*. World Intellectual property organization.

Triola, M. F. (2009). *Estadistica, Decima Edicion*. Pearson,Addison Wesley.

Vaculoy, Alpha Solder. (2022). *Technical Reference Data Sheet for Solder SAC 305.* North America: Mermaid Trademark.

Vitronics Soltec Company. (2011). Technical Data for Delta 5, Vitronics. China: Suzhou.

Well, P. C. B. (2007). *2007-2021 WellPCB Technology Co., Ltd*. PlacaPCB. https://placapcb.com/OSP-PCB-Surface-Finish.html

Yen, Y. W., Chou, W. T., Tseng, Y. U., Lee, C., & Hsu, C. L. (2008). Investigation of dissolution behavior of metallic substrates and intermetallic compound in molten lead-free solders. *Journal of Electronic Materials*, *37*(1), 73–83. doi:10.100711664-007-0266-6

Chapter 10
Implementing Cyber-Physical Systems in Manufacturing Systems

Rithin B. Nambiar
Bosch Automotive Electronics India Pvt. Ltd., India

C. P. Dheeshith
ATKINS, India

K. Abhijith
Sahyadri College of Engineering and Management, India

A. Shahaas
Sahyadri College of Engineering and Management, India

Rathishchandra Ramachandra Gatti
https://orcid.org/0000-0002-4086-2778
Sahyadri College of Engineering and Management, India

ABSTRACT

By employing a constant flow of information from the production processes, manufacturing cyber-physical systems (CPS) is a step ahead of industrial automation from a traditional manufacturing system to a well-networked and adaptable manufacturing system. By learning from the data obtained from real-time and non-real-time production processes that can operate independently, this system can optimize itself and adapt to the changes. With the advent of IoT and CPS, a connection between machine to machine and from machine to human has been established in industrial communication systems, resulting in more efficient and flexible systems that

DOI: 10.4018/978-1-6684-8088-5.ch010

have decreased production time and improved forecasting capacities. CPS focuses on how physical systems and computing networks can communicate. It is applied to IIoT, cyber security, and smart manufacturing. This chapter aims to provide an overview of CPS deployment in manufacturing. It discusses CPS development, CPS adoption, and CPS requirements for Industry 5.0. Finally, the main CPS production problems are also discussed.

1 INTRODUCTION

Cyber-Physical Systems (CPS) is a new development in smart manufacturing that involves computing entities that are closely connected with the surrounding physical world and the continuous interconnected processes between cyberinfrastructure and the physical systems by supplying and accessing data from sensors and communication systems (Wan, Chen, et al., 2013), (Liu & Wang, 2020), (G. Chen et al., 2020). This system can be monitored and controlled by computer-based algorithms connected to the internet and its users, allowing the system to make decisions and operate independently(Tarallo et al., 2018). It has many applications, including transportation, aviation, and defence. CPS is also used in manufacturing automation, connecting machines with the internet for contact-less operations. Integration of CPS in manufacturing is generally known as Cyber-Physical Production System (CPPS) (Herwan et al., 2018), while the term Smart Manufacturing is commercially used. Smart manufacturing uses cutting-edge information and technological developments in manufacturing to build adaptable manufacturing processes and respond to the rapid market requirements. Cyber-physical systems are a crucial smart manufacturing technology widely implemented in designing, operations, and shop floor control(Ding et al., 2020).

The current trends in manufacturing have significant differences from traditional manufacturing, and the current systems have developed in all aspects compared to the practices involved in the first industrial revolution. Traditional manufacturing methods developed during the mass production era mainly focused on economies of scale and machine utilization. They built large inventory units to ensure customer satisfaction and ran the machines for long hours. This approach was inefficient, and thus many changes took place in the industry. The latest iteration in this process is Industry 5.0, characterized by a network of interconnected, automated systems that employ CPS and the Industrial Internet of Things (IIoT).

2 INDUSTRIAL REVOLUTIONS AND EVOLUTION OF CPS

The first industrial revolution began in the late 18th century and was characterized by the mechanization of manufacturing processes, resulting in increased efficiency and productivity. The second industrial revolution, which began in the late 19th century, saw the introduction of mass production techniques and the use of electricity to power machines. The third industrial revolution, also known as the Digital Revolution, emerged in the mid-20th century with the development of computers and automation technologies, leading to significant improvements in manufacturing efficiency. The fourth industrial revolution, or Industry 4.0, is marked by the integration of physical and digital technologies, resulting in smart factories and increased automation. The fifth industrial revolution, which is still in its early stages, is expected to see the rise of new technologies such as quantum computing and biotechnology, which could transform the manufacturing sector and lead to significant advances in fields such as renewable energy and space exploration (George & George, 2020). Despite the significant changes brought about by these revolutions, the manufacturing sector remains a vital part of the global economy, supporting jobs and driving innovation.

2.1 Industry 1.0

Industry 1.0, also known as the first industrial revolution, refers to the era between the late 18th century and the early 19th century, during which substantial technical improvements revolutionised the production of products and contributed to the development of the modern industrial economy. This period is defined by the move from manual labour to machine-based manufacturing, making it possible to produce many bigger-scale items. The creation of the steam engine was the primary technical advancement of Industry 1.0(Bigliardi et al., 2020). Initially, steam engines were employed to pump water out of coal mines, but they quickly became integral to early industries. By harnessing the force of steam, industries could run machinery much more productive than human labour. Technology enabled the automation of the manufacture of commodities, which eventually led to improved productivity, decreased costs, and a rise in economic development. The steam engine's invention contributed to the textile industry's development (Spear, 2008). The textile sector was one of the first to embrace new technology, and it became a key force in the Industrial Revolution almost immediately. With machinery such as the spinning jenny and the power loom, creating textiles on a huge scale and at a lower price than ever was possible. Additional developments of Industry 1.0 included the creation of new iron-making techniques, such as the Bessemer process, which made mass production of steel conceivable. This resulted in the rise of the railroad business and the building of enormous bridges and structures. Overall, Industry 1.0 was a time of significant technological innovation that revolutionised the production of products and established the basis for the contemporary industrial economy (Berg, 2005). As the world continues to depend on automation and modern production processes, this period's developments continue to influence society.

2.2 Industry 2.0

Industry 2.0, often known as the second industrial revolution, refers to the mid-19th century to the early 20th-century era of rapid industrialisation and technological growth. It was distinguished by new inventions that built upon the successes of Industry 1.0, such as the introduction of mass manufacturing processes, new communication technologies, and new forms of energy generation. The introduction of the assembly line was one of the most important advances of Industry 2.0 (Yin et al., 2018). Henry Ford pioneered this technique, making mass manufacturing of things far quicker. Each worker would concentrate on a specialised activity, with the product travelling down a conveyor belt from one worker to the next. This significantly enhanced productivity and decreased manufacturing costs, resulting in the widespread availability of inexpensive consumer products. Another significant advance of Industry 2.0 was the development of new energy generation techniques. The steam engine was still in use, but other technologies, like the internal combustion engine and electricity (Sharma & Singh, 2020), quickly replaced it. The invention of the internal combustion engine spawned the automotive industry, whereas the broad adoption of electric lights and powered machines resulted from the availability of electricity. In addition, developing new communication technologies was a crucial component of Industry 2.0. The telegraph was created during Industry 1.0, whereas the telephone was invented during Industry 2.0. The telephone changed communication and made instantaneous long-distance communication feasible. The invention of the radio also had a significant effect, allowing for the widespread diffusion of knowledge and the emergence of a new type of entertainment. Industry 2.0 was a time of significant change in the production and communication of commodities. It paved the way for the contemporary industrial economy and established the groundwork for several technical advancements we depend on today.

2.3 Industry 3.0

Industry 3.0, sometimes known as the third industrial revolution, refers to the late 20th-century era of technological development. It was marked by the advent of digital technology, which permitted automation and the development of intelligent industries. The emergence of computers and the internet ushered in a new age of interconnection and digitalisation, radically altering product production and distribution. The introduction of computer-controlled production was one of the fundamental advances of Industry 3.0. This allowed for more accuracy and control in the manufacturing process and the capacity to make immediate production modifications in response to demand fluctuations. This technology also allowed for the development of "smart" factories, where machines and equipment could interact with one another to maximise productivity. The general use of the internet during Industry 3.0 was an additional key breakthrough (Luca & Dolga, 2021). This enabled the emergence of new kinds of communication, such as email and instant messaging, and e-commerce platforms that revolutionised the distribution and sale of commodities. The advent of the internet also facilitated the development of new business models, such as software as a service (SaaS) and cloud computing, which have become indispensable to current company operations. In addition, new technologies like robots and Artificial Intelligence (AI) were created due to the emergence of Industry 3.0 (Tantawi et al., 2019). The combination of robotics and AI-enhanced the capabilities of automated systems even more. Industry 3.0 was a time of substantial product production, distribution, and consumption change. The emergence of digital technology permitted increased efficiency and productivity and created new corporate and commercial opportunities. Industry 3.0 advancements continue to change how we live and work today. The precursors of CPS were developed in this industrial revolution.

2.4 Industry 4.0

Industry 4.0, the fourth industrial revolution, refers to technological upheaval in production and industry. It is defined by incorporating modern technologies such as the Internet of Things (IoT), AI, and advanced robotics, which build on the advancements of Industry 3.0. The utilisation of Cyber-Physical Systems (CPS), which link physical things and machines to the internet and other digital systems, is one of the defining characteristics of Industry 4.0 (Pivoto et al., 2021). This allows for real-time monitoring and management of the manufacturing process and the collection and analysis of huge quantities of data. This information may be utilised to improve production, discover inefficiencies, and change the manufacturing process in real-time. AI and machine learning is an additional significant part of Industry 4.0. These technologies allow machines to study and learn from data, increasing their efficiency and effectiveness. AI may also be utilised to develop autonomously deciding and acting intelligent systems, boosting automated systems' capabilities. Industry 4.0 is also characterised by its reliance on robots. Sophisticated robots provide for more automation and flexibility in production and the capacity to accomplish complicated tasks that formerly required human interaction. Cobots, or collaborative robots, may operate alongside human operators and help them complete tasks, enhancing productivity and safety (Matheson et al., 2019).

Moreover, modern materials and additive manufacturing methods, such as 3D printing, are included in Industry 4.0. These technologies may minimise waste and the environmental effect of manufacturing while enabling the manufacture of complex and customised components and products. Overall, Industry 4.0 signifies a significant change in the production and manufacturing of commodities. This era of fast technological growth has the potential to disrupt industry and production profoundly. In the next years, it is anticipated that the innovations of Industry 4.0 will substantially influence the economy, society, and environment.

2.5 Industry 5.0

Experts in manufacturing and industry are now debating the futuristic notion of Industry 5.0. It is sometimes referred to as the next phase of industrial development, building on the breakthroughs of Industry 4.0 while emphasising human-centred production and sustainability. One of the core concepts of Industry 5.0 is that technology should be utilised to augment rather than replace human skills. This implies that modern technology such as AI, robots, and automation should be intended to complement human labour rather than replace it. This may lead to a more collaborative and adaptable production process in which people and machines collaborate to produce high-quality goods. Industry 5.0 is also notable for its emphasis on sustainability and the environment. Industry 5.0's primary concept is a circular economy, where waste is reduced, and resources are reused and repurposed. Industrial processes should be designed to decrease waste and pollution and, if feasible, use renewable energy sources.

Moreover, Industry 5.0 highlights the significance of customisation and individualisation. With cutting-edge technology such as AI and machine learning, it will be able to design goods that are personalised to each client's unique requirements and tastes. This might result in increased consumer satisfaction and loyalty. Industry 5.0 also includes the usage of digital twins or virtual versions of real things and systems (Jagatheesaperumal & Rahouti, 2022). Digital twins may mimic and improve manufacturing processes, enabling more rapid and effective output. In addition, they may be used to monitor the functioning of items in real-time, allowing for predictive maintenance and decreasing downtime. Industry 5.0 is a vision for the future of manufacturing and industry that is human-centred, environmentally conscious, and technologically sophisticated. Several experts feel that the advances of Industry 4.0 are setting the groundwork for the future growth of Industry 5.0, even though this is still a theoretical idea.

2.6 Industry 6.0

Industry 6.0 would build upon the achievements of Industry 5.0 and propel manufacturing and industry to new heights. Industry 6.0 is currently a theoretical notion and has yet to be a commonly recognised phrase. Although the precise characteristics of Industry 6.0 are unknown, there are various hypotheses on what it may involve. Advanced bioengineering and biomanufacturing methods are a possible feature of Industry 6.0. This might lead to a new age of biomanufacturing, including using live cells and creatures to manufacture new materials and products. This might generate new, sustainable materials or previously impossible-to-manufacture items. Integrating virtual and augmented reality (VR/AR) technology into production might be another part of Industry 6.0 (Duggal et al., 2022). This might enable more immersive and interactive design and manufacturing processes and improve worker training and cooperation. Fully automated manufacturing processes are an additional potential for Industry 6.0. This might entail using sophisticated robots and AI to construct industrial systems that manage themselves and need little human participation. This might result in extremely efficient and simplified production processes and substantial cost reductions. Industry 6.0 is also defined by the growing prominence of nanotechnology, which has led to novel materials with exceptional qualities. Nanotechnology is the manipulation of materials on the size of atoms and molecules, creating novel materials with unique features and possible industrial uses. Although the precise characteristics of Industry 6.0 remain unknown, technological advances will certainly continue to substantially impact the future of manufacturing and industry. The breakthroughs of Industry 6.0 have the potential to usher in a new age of manufacturing in which modern technology and novel materials are used to build previously inconceivable items. The CPS of this industrial revolution is considered to be very advanced with very close integration of cyber and real-world spaces.

2.7 Evolution of CPS in These Five Industry Revolutions

Cyber-physical systems (CPS) have evolved significantly during these five industrial revolutions. In the first industrial revolution, CPS did not exist as a concept. The second industrial revolution saw the introduction of electrical and mechanical systems, which could be considered precursors to modern CPS. However, it was not until the third industrial revolution that CPS emerged as a distinct concept, with the development of computer-controlled manufacturing systems. The fourth industrial revolution, or Industry 4.0, is characterized by the integration of physical and digital technologies, including CPS. Smart factories use sensors, actuators, and other devices to collect and analyze data, allowing for real-time optimization and increased efficiency (Frontoni et al., 2018). CPS are also used in fields such as transportation, healthcare, and agriculture. The fifth industrial revolution is expected to see the continued evolution of CPS, with the integration of new technologies such as quantum computing and biotechnology. This could lead to significant advances in fields such as precision medicine and autonomous transportation.

3 ABOUT CPS AND ITS GENERAL ARCHITECTURE

CPS is a network connection between computational and physical systems that enables the system to control the complex coordination of multiple systems (Wan, Yan, et al., 2013). Sensors and other embedded systems are used for data acquisition from the real world, the physical system. The signals from the sensors are analogous, and it is thus converted to digital using analog to digital converters (ADC). It is then sent to cyberspace through communication networks—data encryption and integration safeguard data privacy and integrity, which are very important in a network. The CPS system should ensure the security and reliability of data received by the end node (Institute of Electrical and Electronics Engineers, n.d.).

A cyber system consists of a control center and a data center. Data sent through sensor networks are stored at the data center. Before giving any instructions to the control center, the credibility of the data will be checked by the data center. The control center will send control commands to the actuator networks after receiving instructions from the data center. The actuator network will send the needed instructions to the associated nodes.

Figure 1. Basic structure of CPS

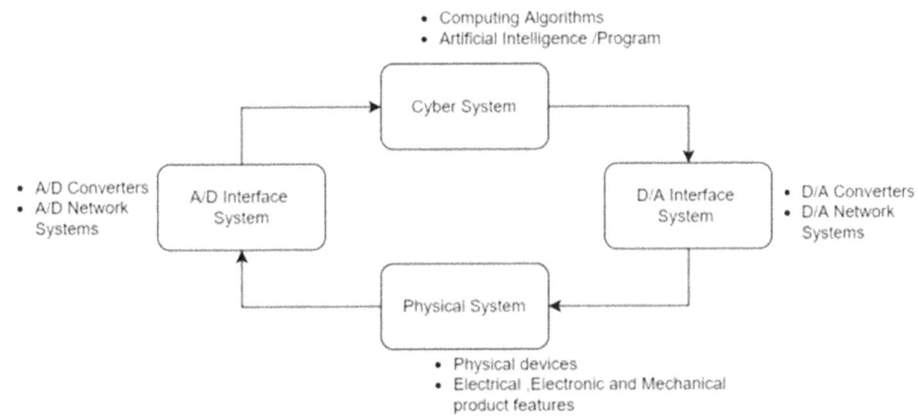

The cyber system consists of computational algorithms and programs that support the discretion of the CPS operations. The algorithms may be instruction-based programs for performing basic tasks with little or no feedback or AI-based programs(Radanliev et al., 2021) with frequent feedback mechanisms. They may be specific AI or Artificial General Intelligence (AGI) if they are based on AI algorithms. Most currently developed systems are in the specific AI, but the trend is towards adding more generic features to the current CPS to transition to AGI.

The physical system consists of the tangible product, process, or plant that needs to be controlled using the cyber system. These include the physical devices and product features that form the controllable physical entities. Typically, they consist of sensors, actuators, microcontrollers, microprocessors, amplifiers, and mechanisms, to name of few.

The interface systems consist of communication devices between the cyber and physical systems. There are two interface systems – Analog to Digital (A/D) Systems for communication from a physical system to a cyber system and Digital to Analog (D/A) systems for communication from a cyber system to the physical system. Typical interface systems are A/D converters and D/A converters.

4 APPLICATIONS OF CPS

The rapid development of CPS technology has supported many other key technologies and product development in automobile, defence, industrial automation, medical equipment, intelligent transport systems, control systems, and many other industries. CPS in the field of medicine has applications in medical equipment and services. The massive amount of data collected from hospitals and pharmacy need to be stored in cloud storage that can be accessed remotely. CPS is also employed in robotic surgeries and for specialist doctors to remotely monitor the patient's health condition and drug administration. CPS has its involvement in the generation, transmission, and distribution of electricity in smart grid technology. It also helps in the communication of consumers with the smart grid system. CPS provides a solution for traffic congestion, pollution, and safety on the road. An intelligent transportation system enables feedback-back-based interactions between the physical and cyber systems. Unmanned vehicles and autopilot are also driven with the support of CPS. In precision agriculture, CPS helps in the decision-making operations by collecting and computing the fertility and humidity of the soil and with automated irrigation. However, one of the most significant domains where CPS can be readily applied in manufacturing.

CPS in manufacturing helps remotely communicate between machine and machine operator. With the introduction of IIoT and high-speed networks, manufacturing has become a process based on a continuous feedback system for smart manufacturing and big data. Failure in the system is detected early and with predictive maintenance, the time and cost of production are reduced.

5 STATE OF THE ART MANUFACTURING CPS

Cyber-physical systems play an essential role in the field of smart manufacturing. During the Industry 2.0 era, most manufacturing industries began to use mechanical systems in the production lines. The beginning of the 20th century marked a tremendous increase in the demand for products that mechanical systems alone cannot produce to meet the requirements. So electronic systems and information systems were introduced into the production lines leading to automation and intelligence in manufacturing processes.

During the industry 4.0 era, the internet is already in every field. Technical advancements also created connections among people, things, and spaces using the internet. Internet not only connected humans to humans but also established connections between humans to machines and even machines to machines. Beyond the IoT, intelligent IoE (Internet of Everything) became possible as a connection between the real world and cyberspace (Kim & Park, 2017). Due to these successive industrial revolutions, data transmission technologies have evolved from mechatronics to more advanced cyber-physical systems (1.0 to 5.0). The integration of mechanical systems with electrical, electronics, computer, and control engineering is termed mechatronics. This system is costly, and there were delays in processing due to a lack of understanding across the different disciplines. CPS was introduced as a solution to these problems.

CPS is a concept that combines physical processes with computing and provides modelling, abstractions, design, and analysis approaches for the combined system. The cyber-physical system consists mainly of sensors, actuators, and controllers. Furthermore, these elements helped reduce the operation cost and optimize the time taken for data processing. The CPS technology has wide applications in digital medical instruments, aerospace control technology, and smart manufacturing systems.

6 IMPLEMENTATION OF MANUFACTURING CPS

CPS plays a vital role in the manufacturing process. CPS has an essential role, from monitoring the physical systems and the machines to creating a digital model of the physical system. CPS enables the system to be available and accessible for humans to operate and control from any location. In CPS-enabled manufacturing systems, the machines are connected with IIoT to communicate with human beings through Human Machine Interface (HMI) technologies and from Machine-to-Machine(M2M) communication using the internet (Ahmadi et al., 2019). Thus, in a manufacturing CPS, a collaborative system is developed between different systems, including the sensors installed in the physical system, wired or wireless network, computational systems, cloud, and internet. The construction of CPS is getting complex with the increasing use of IIoT, access to communication, big data, and network connectivity(Comput Sci Inf Technol, n.d.-a).

6.1 Physical System in Manufacturing CPS

The physical system in CPS manufacturing consists of machines, sensors, actuators, computing devices, and tracking elements. The real-time data collected from the sensors integrated with the physical systems or machines are sent to the computing devices and cloud systems through wired and wireless networks. Further processing of the data is completed in the cyber system, and actuators carry the operations in the physical systems from the signal they receive from the computing devices(Al-Ali et al., 2018).

Sensors are the primary elements in a physical system. The output signal of the sensors may not be compatible with the input of the computing elements or the controllers, thus requiring intermediary signal processing technologies. However, some standardized sensors can convert machines' mechanical motions and other analog signals into digital or electrical signals compatible with the controllers. These electrical signals are sent to the computation device through networks. Sometimes these signals also require to be processed through signal conditioners to maintain the level of signals. The network can be wired or wireless. These networks use specific communication protocols for the transfer of data. The distance to which the data or signals can be carried depends on the communication protocols. The data from the sensors are processed under the control algorithm by the controllers. After computation, the resultant signal is sent back to the actuators for actuation or to the storage space. The actuators or other elements in the physical system respond to the digital signals. Actuators play a vital role in the CPS physical system. It changes the system's status under production with the commands from the controller. Power amplifiers and relays are employed in the circuit to generate the necessary power for actuation. The system is actuated to meet the operational requirements needed during the manufacturing process.

6.2 Cyber System

The cyber system acts as an information center in the manufacturing CPS architecture. Data or information is transmitted to the cyber system from every machine and back. Analytics is used to extract additional information from the collected massive data. The cyber system in manufacturing CPS consists of the computing system, storage, and communication networks. The data from the physical systems are analyzed and processed in the computation system, which can be a physical system such as a microcontroller or a system in cyberspace. The information from the machines is transferred to the cyber computation system through networks. Thus it can be accessed via wired or wireless networks. The data is stored in cloud storage which can be accessed from anywhere. More data can be stored and accessed through big data. It also enables the storage of a massive amount of data related to the machines and operations. The machine can access this data whenever required, which makes it self-sufficient and helps in automation. Cloud computing allows the handling of a massive amount of data and the modelling of various industrial procedures (Moeuf et al., 2018). The manufacturing CPS system depends on sharing data or information among manufacturing centers and logistics departments. Most industries use the internet to connect with other manufacturing units or between devices. The operator can access the data and live status of machines and control them remotely. He can also fix specific problems within the machines remotely and thus reduce the idle time of machines—data analysis and machine learning help to reduce the cost of operation and production by optimizing the production.

7 COMMUNICATION PROTOCOLS AND ELEMENTS IN MANUFACTURING CPS

The manufacturing CPS have access to cyberspace with different networking protocols. Wi-Fi, WiMAX, and cellular technology are some of the widely used network protocols. Industry 4.0 uses Industry Wireless Networks (IWNs) to transmit data and control. Industries use big data, industrial cloud, and industrial equipment to increase productivity and reduce energy and cost. In industry, the data being communicated includes personal data of customers, important information on factories and services, and manufacturing. Quality of service (QoS) of IWNs, which includes real-time security, reliability, and longevity, is a significant concern in data communication in the industry. Wireless communication increases the degree of flexibility, usability, versatility, and efficiency in CPS manufacturing (Ahmadi et al., 2018).

Nevertheless, many industrial applications have high requirements for reliability and real-time capability. The required wireless network for a manufacturing CPS is selected considering its requirements. IoT plays an essential role in M2M communication in Industry 4.0. CPS and IIoT-connected devices are located physically in different geographical regions and industries for communication and processing. Wireless communication networks can be controlled remotely. It helps ease deployment and covers a wide area for production and sales. Recent news can operate in challenging industrial environments such as high temperature and vibration (Institut de technologie de Cambodge et al., n.d.), (H. Chen, 2017).

Sensing, communication, computing, and control are integrated with CPS by supporting communication networks. The flexibility of CPS allows the designer to select the required communication technology. Only a single type of communication may not be sufficient. Thus, different standards of communication protocols are integrated into the manufacturing CPS. Understanding and choosing the correct industry-standard protocols is crucial, depending on the custom Industry 4.0 requirements. Some communication protocols used in industries are Zigbee, Bluetooth, Wi-Fi, LoRaWAN, PON, and cellular networks such as 4G/5G (Iqbal et al., 2020).

7.1 Zigbee

Zigbee is a wireless networking technology developed to reduce a communication network's cost and power consumption. It also enables seamless M2M communications and IIoT communication networks. It helps in the automation control and communication between sensors and computation networks. Zigbee can be widely adopted in the industry since it can contain up to 65000 nodes and connect them in any industry as a single control network (Shetty et al., 2021).

7.2 Bluetooth

Short-range wireless communication is enabled with Bluetooth networks. It has significantly less power consumption while providing a wide connectivity range (Mahmoud & Mohamad, 2016). Bluetooth is progressed to provide high speed and greater flexibility. From a traditional usage of audio streaming, Bluetooth has now developed into applications like mesh connectivity for M2M communications, connectivity of CPS, and networking of IIoT. It is used for no-streaming sensor node-type applications.

7.3 Wi-Fi

Wi-Fi (Wireless Fidelity) is a widespread communication protocol that offers high-speed data transmission and works optimum for wireless local area network environments. The growing industrial requirements are met by releasing updated variants that have increased data transfer speed and carry more data. Infrastructure and ad-hoc networks are two types of WLAN networks employed in the industries. Infrastructure networks use devices such as routers to connect machines or equipment, while ad-hoc network works without the connected devices. Wi-Fi, along with other communication standards, is making its way into the market with increasing speed and range of data transfer.

7.4 LoRaWAN

LoRaWAN is a networking protocol that consumes less power and has a wide networking area connecting battery-operated devices wirelessly. It provides a cellular style of the communication network. Furthermore, it is mainly used in IoT networks and machine-to-machine communication.

7.5 Passive Optical Network (PON)

This telecommunication system brings optical fiber cabling and signals to the end-user. The optical line terminates at the required location. In the case of the industry, the end-user will be the industrial equipment. The optical network from the network center of the industry is split into different types of equipment as per the requirements. Optical data transmission consumes no or less power and does not require active electronic devices between the network (Tucker, 2010).

7.6 Cellular Networks

Cellular network technology is a widely used network protocol that connects different devices based on frequency requirements. The functionality of different generations of the cellular network has advanced with the generations. In industrial applications, mainly a 4G or 5G network is used because of its extensive coverage range, less power consumption, bandwidth, and spectrum. It enables a seamless operation of devices that move around the shop floor with the wireless network. The speed of data transfer is also more than other wireless communication protocols. Because of the advantages of cellular networks, industrial enterprises are moving towards mobile operators as partners to provide private networks.

8 REQUIREMENTS OF MANUFACTURING CPS IN INDUSTRY 5.0

The primary objective of Industry 5.0 is to develop a more evolutionary and progressive evolution of Industry 4.0. Cobots, or collaborative robots, are a concept introduced in Industry 5.0. Cobots are tools required to meet today's needs for enterprises that offer customized products for customers. Human intelligence is brought into the picture by collaborative robots. In the fifth industrial revolution, the CPS consisted of humans, artificial intelligence, and the physical systems of enterprises linked via high-speed internet (Pathak et al., 2019). Devices that function in partnership with control and computation devices are used to achieve communication between various components of a CPS. Manufacturing industries are now enhancing their strength with human-robot collaboration, allowing them to adapt to rapidly changing market requirements. The fifth revolution comprises the combination of advanced technologies and humans, with robots completing the manufacturing and IIoT devices collecting data necessary for production. Companies employ AI software to learn and understand specific procedures to input and evaluate data with other systems.

8.1 Virtual Manufacturing

Virtual manufacturing is defined as manufacturing computers. Traditional manufacturing is carried out under controlled conditions and concentrates on machinery, labor, material, money, and knowledge. The scenario in current manufacturing technologies has advanced with CPS. Anyone can easily collect knowledge and information anywhere outside the factories, even from a personal computer with the internet and multimedia. With CAD, CAM, and CAE software, one can virtually design a product or process with customized design. The real-time physical structure can be understood and simulated from this virtual design. The design can also be analyzed under different conditions, and required changes can be made quickly. This will reduce the manufacturing cost and also the production time.

8.2 Digital Twin

A digital twin in manufacturing is the digital replica for the virtual representation of a real-world component used in the production process (Nikolakis et al., 2019). The digital representation utilizes inputs from a real-world component. The digital twin mirrors the real component's status, functionality, and interaction with other devices. Digital twin technology has a wide range of applications in the manufacturing, automotive, agriculture, healthcare, and aerospace industry. A digital twin is employed in the manufacturing industry for various operations such as improving system design, testing new products, monitoring and maintenance, and Product Lifecycle Management (PLM), which can be done remotely with the help of CPS. The real-time data from the machinery is collected, such as vibration, humidity, motion, and the maintenance team will be informed before failure. Instead of a lengthy process and trial and error method of testing, digital twin experts can simulate different scenarios faster and easier, lowering the cost. The manufacturing team can use the digital twin to plan and test new production lines. Warehouse designs can be planned more quickly in this way. With the digital twin technology, manufacturers can provide AR (Augmented Reality) programs to the maintenance department. Technicians can observe the up-to-date model of the product with AR glasses. Altogether the models help in advanced manufacturing (Catarci et al., 2019).

8.3 Collaborative Human-Robot Interactions

Industry 5.0 is basically about the collaborative working environment between humans and robots. It is a state of development where skilled labor and robots' ability to join together in a manufacturing process. The involvement of human labor will be brought back in industry 5.0. Robotic abilities and human intelligence will increase the productivity and efficiency of the manufacturing system. With the collaboration of robots, manufacturing is completed at a higher speed and efficiency. This system requires the labor to use language understandable to the robot. CPS enables remote human-robot collaboration (Wang et al., 2020). Human safety from uncontrolled robotic actions is essential in a collaborative environment. Establishing a closed-loop control system based on the person's proximity to the robot allows for human safety. The collision risk detection module has been installed and is responsible for initiating the proper collision avoidance procedure. The vision and distributed computing systems enable low processing and evaluation times, enabling real-time safety assessment and collision avoidance tactics. The human can make the robot work with hand gestures as well. Communication with robot and safety monitoring is made accessible using CPS.

8.4 Blockchain

Blockchain is a foundation for a secured communication protocol in the manufacturing CPS. This technology enables users to create a decentralized peer-to-peer network with data security. Blockchain can track any changes in the data and secure data without third-party authentication. Each network node has its own shared distributed database, which comprises a series of information blocks, with each block linked to the next using a unique cryptographic identity. An agreement between each mode is required to add a new block to the network. For this, the nodes use smart contract technology. Blockchain enables a decentralized and secured system with transparent applications. This technology is coupled with smart contracts, and the system's security is enhanced by removing the reliance on third parties. In Manufacturing CPS, cloud services have more access, and blockchain technology is applied to improve data security (Vatankhah Barenji et al., 2020).

9 CHALLENGES IN IMPLEMENTATION OF MANUFACTURING CPS

In an industry 5.0 environment, CPS and IIoT are becoming increasingly important. Many intelligent sensors are now available, resulting in a large amount of data. Several issues arrive with big data in the industry. This includes a vast network connection with security threats, access protocol issues, and the quality of the data collected from the industries. An efficient data storage system and cloud services with cyber protection are inevitable for manufacturing operations controlled by CPS (Comput Sci Inf Technol, n.d.-b).

9.1 Data Security of Manufacturing CPS

The manufacturing industry is one of the primarily targeted industries for cybercrime. The manufacturing industry is attacked to steal the client information from the customer resource management system and hold it for ransom. Hackers also hack into the system to take control of the manufacturing process and tamper with the products. Data is also being stolen for the advantage of competitors. The storage of confidential information is essential. When it comes to CPS and IIoT, the information can be accessed in more ways than before, as most industries nowadays depend on cyber systems and cloud storage. With proper time and financial investment in cyber security, the industry can enjoy all the benefits of manufacturing CPS.

9.2 Big Data and Data Handling of Manufacturing CPS

The CPS-based manufacturing system continuously collects data from the sensors and stores the data for future needs. Thus, a vast amount of data is collected in the cyber system. Instead of collecting random data and filtering them for the extraction of useful information, only the required data with priority is being collected and stored. Some information is sent to the database, requiring immediate response after computation. The computation speed and network speed should match the requirements. The computation also requires more memory and faster CPUs, and this hardware and software are expensive. This data is exposed to cyber threats like malware and denial of services. Attackers exploit vulnerabilities in database management systems and make the system unavailable to the users. Many smaller enterprises lack expert knowledge in the application of cyber security. This always makes way for security breaches. The companies can invest in cyber security and train the employees to protect the data storage system (Lidong & Guanghui, 2016).

9.3 High-Speed Communication Networks

Manufacturing enterprises that depend on CPS require an uninterrupted high-speed internet connection. A high-speed network is required for the efficient operation and reliability of the system. The emerging 5G technology will enable operators to have high-speed communication between factories and machines (Rao & Prasad, 2018). The network inside the factory can be Wi-Fi or Bluetooth, which does not require an extended range of connections. When the control is required for long-range, a cellular network is used. The production rate and reliability of the manufacturing system depend on the speed and range of the network connection. The high-speed network connection should be maintained as long as the factory operates (O'Connell et al., 2020).

9.4 Managerial Challenges

A high initial investment is required to set up a cyber-physical system-based manufacturing industry. The change from the traditional manufacturing plant to an advanced and intelligent factory demands higher efforts and costs for the planning and implementation. The management's lack of knowledge and experience in coordinating the machines and supply chain makes the system inconvenient. Managers should take the initiative to enhance awareness among others on the deployment of CPS in the manufacturing industry. Every industry should also encourage high-end research activities on the industry's future development with industry 5.0.

9.5 Reliability Challenges in CPS

Since the Manufacturing CPS relies heavily on sensor-based data for decision making, the sensors and sensor nodes of Wireless Sensor Networks (WSNs) tend to be less reliable, especially if they are battery-powered. New technologies such as energy harvesting can be employed to make these ubiquitously distributed sensors self-powered, thus increasing their reliability. Also, since manufacturing CPS gets increasingly complex with more physical infrastructure such as machines, sensors, and actuators, reliability becomes more challenging (Gunes et al., 2014).

10 CONCLUSION

An overview of CPS in manufacturing is discussed in this paper. CPS has its inevitable role in industry 4.0, where cyber systems monitor and control most manufacturing operations. The physical system consists of sensors and machines. In Industry 4.0, a majority of human labor was replaced by cyber systems. Industry 5.0 brings back more human involvement in manufacturing operations. With industry 5.0, humans and robots started to work together for an increased production rate and process optimization. The reliability of CPS manufacturing systems depends on the communication networks where the physical and cyber systems are connected and communicated with high-speed data transfer protocols. The development of virtual manufacturing and digital twin enables the manufacturing and maintenance from remote locations with CPS (Tao et al., 2019). The main challenges for CPS in industry 4.0 and 5.0 are big data, data privacy, cybersecurity, and an uninterrupted high-speed network. Blockchain technology enables the security of the data by encrypting the data shared between two nodes. The speed of communication networks has increased with the 5G technology. Many new technologies are expected in the CPS manufacturing systems with the development of new cyber security measures and a 5G cellular network.

REFERENCES

Ahmadi, A., Moradi, M., Cherifi, C., Cheutet, V., & Ouzrout, Y. (2018). Wireless connectivity of CPS for smart manufacturing: A survey. *2018 12th International Conference on Software, Knowledge, Information Management & Applications (SKIMA)*, (pp. 1–8). ACM.

Ahmadi, A., Sodhro, A. H., Cherifi, C., Cheutet, V., & Ouzrout, Y. (2019). Evolution of 3C cyber-physical systems architecture for industry 4.0. *Service Orientation in Holonic and Multi-Agent Manufacturing: Proceedings of SOHOMA 2018*, (pp. 448–459). ACM.

Al-Ali, A. R., Gupta, R., & al Nabulsi, A. (2018). Cyber physical systems role in manufacturing technologies. *AIP Conference Proceedings*, *1957*, 050007. doi:10.1063/1.5034337

Berg, M. (2005). *The age of manufactures, 1700-1820: Industry, innovation and work in Britain*. Routledge. doi:10.4324/9780203990971

Bigliardi, B., Bottani, E., & Casella, G. (2020). Enabling technologies, application areas and impact of industry 4.0: A bibliographic analysis. *Procedia Manufacturing*, *42*, 322–326. doi:10.1016/j.promfg.2020.02.086

Catarci, T., Firmani, D., Leotta, F., Mandreoli, F., Mecella, M., & Sapio, F. (2019). A conceptual architecture and model for smart manufacturing relying on service-based digital twins. *Proceedings - 2019 IEEE International Conference on Web Services, ICWS 2019 - Part of the 2019 IEEE World Congress on Services*, (pp. 229–236). IEEE. 10.1109/ICWS.2019.00047

Chen, G., Wang, P., Feng, B., Li, Y., & Liu, D. (2020). The framework design of smart factory in discrete manufacturing industry based on cyber-physical system. *International Journal of Computer Integrated Manufacturing*, *33*(1), 79–101. doi: 10.1080/0951192X.2019.1699254

Chen, H. (2017). Theoretical Foundations for Cyber-Physical Systems: A Literature Review. *Journal of Industrial Integration and Management*, *02*(03), 1750013. doi:10.1142/S2424862217500130

Ding, K., Lei, J., Zhang, F., Wang, Y., & Wang, C. (2020). Analyzing the cyber-physical system–based autonomous collaborations among smart manufacturing resources in a smart shop floor. *Proceedings of the Institution of Mechanical Engineers. Part B, Journal of Engineering Manufacture*, *234*(3), 489–500. doi:10.1177/0954405419875340

Duggal, A. S., Malik, P. K., Gehlot, A., Singh, R., Gaba, G. S., Masud, M., & Al-Amri, J. F. (2022). A sequential roadmap to Industry 6.0: Exploring future manufacturing trends. *IET Communications*, *16*(5), 521–531. doi:10.1049/cmu2.12284

Frontoni, E., Loncarski, J., Pierdicca, R., Bernardini, M., & Sasso, M. (2018). Cyber physical systems for industry 4.0: towards real time virtual reality in smart manufacturing. *Augmented Reality, Virtual Reality, and Computer Graphics: 5th International Conference, AVR 2018, Otranto, Italy, June 24–27, 2018. Proceedings*, *5*(Part II), 422–434.

George, A. S., & George, A. H. (2020). Industrial revolution 5.0: The transformation of the modern manufacturing process to enable man and machine to work hand in hand. *Journal of Seybold Report ISSN*, (1533), 9211.

Gunes, V., Peter, S., Givargis, T., & Vahid, F. (2014). A survey on concepts, applications, and challenges in cyber-physical systems. [TIIS]. *KSII Transactions on Internet and Information Systems*, *8*(12), 4242–4268.

Herwan, J., Kano, S., Oleg, R., Sawada, H., & Kasashima, N. (2018). *Cyber-Physical System Architecture for Machining Production Line.*

Institut de technologie de Cambodge, Institute of Electrical and Electronics Engineers. Islamabad Section, & Institute of Electrical and Electronics Engineers. (n.d.). *SKIMA 2018 : the 12th International Conference on Software, Knowledge, Information Management & Applications*. IEEE.

Institute of Electrical and Electronics Engineers. (n.d.). *2018 International Conference on Computing, Electronic and Electrical Engineering (ICE Cube)*. IEEE.

Iqbal, M. A., Hussain, S., Xing, H., & Imran, M. A. (2020). *Enabling the internet of things: fundamentals, design and applications*. John Wiley & Sons.

Jagatheesaperumal, S. K., & Rahouti, M. (2022). Building Digital Twins of Cyber Physical Systems With Metaverse for Industry 5.0 and Beyond. *IT Professional*, *24*(6), 34–40. doi:10.1109/MITP.2022.3225064

Kim, S., & Park, S. (2017). CPS(Cyber Physical System) based Manufacturing System Optimization. *Procedia Computer Science*, *122*, 518–524. doi:10.1016/j.procs.2017.11.401

Lidong, W., & Guanghui, W. (2016). Big Data in Cyber-Physical Systems, Digital Manufacturing and Industry 4.0. *International Journal of Engineering and Manufacturing*, *6*(4), 1–8. doi:10.5815/ijem.2016.04.01

Liu, H., & Wang, L. (2020). Remote human–robot collaboration: A cyber–physical system application for hazard manufacturing environment. *Journal of Manufacturing Systems*, *54*, 24–34. doi:10.1016/j.jmsy.2019.11.001

Luca, D., & Dolga, V.-I. (2021). From Industry 3.0 to Industry 4.0. *The International Journal of Engineering and Science*, *10*(9), 44–49.

Mahmoud, M. S., & Mohamad, A. A. H. (2016). *A study of efficient power consumption wireless communication techniques/modules for internet of things (IoT) applications*. Research Gate.

Matheson, E., Minto, R., Zampieri, E. G. G., Faccio, M., & Rosati, G. (2019). Human–robot collaboration in manufacturing applications: A review. *Robotics (Basel, Switzerland)*, *8*(4), 100. doi:10.3390/robotics8040100

Moeuf, A., Pellerin, R., Lamouri, S., Tamayo-Giraldo, S., & Barbaray, R. (2018). The industrial management of SMEs in the era of Industry 4.0. *International Journal of Production Research*, *56*(3), 1118–1136. doi:10.1080/00207543.2017.1372647

Nikolakis, N., Alexopoulos, K., Xanthakis, E., & Chryssolouris, G. (2019). The digital twin implementation for linking the virtual representation of human-based production tasks to their physical counterpart in the factory-floor. *International Journal of Computer Integrated Manufacturing*, *32*(1), 1–12. doi:10.1080/0951192X.2018.1529430

O'Connell, E., Moore, D., & Newe, T. (2020). Challenges associated with implementing 5G in manufacturing. *Telecom*, *1*(1), 5. doi:10.3390/telecom1010005

Pathak, P., Pal, P. R., Shrivastava, M., & Ora, P. (2019). Fifth revolution: Applied AI & human intelligence with cyber physical systems. *International Journal of Engineering and Advanced Technology*, *8*(3), 23–27.

Pivoto, D. G. S., de Almeida, L. F. F., da Rosa Righi, R., Rodrigues, J. J. P. C., Lugli, A. B., & Alberti, A. M. (2021). Cyber-physical systems architectures for industrial internet of things applications in Industry 4.0: A literature review. *Journal of Manufacturing Systems*, *58*, 176–192. doi:10.1016/j.jmsy.2020.11.017

Radanliev, P., De Roure, D., Van Kleek, M., Santos, O., & Ani, U. (2021). Artificial intelligence in cyber physical systems. *AI & Society*, *36*(3), 783–796. doi:10.100700146-020-01049-0 PMID:32874020

Rao, S. K., & Prasad, R. (2018). Impact of 5G technologies on industry 4.0. *Wireless Personal Communications*, *100*(1), 145–159. doi:10.100711277-018-5615-7

Sharma, A., & Singh, B. J. (2020). Evolution of industrial revolutions: A review. *International Journal of Innovative Technology and Exploring Engineering*, *9*(11), 66–73. doi:10.35940/ijitee.I7144.0991120

Shetty, S. H., Rao, A., & Gatti, R. R. (2021). *State of the Art Review of IIoT Communication Protocols*., doi:10.4018/978-1-7998-7541-3.ch003

Spear, B. (2008). James Watt: The steam engine and the commercialization of patents. *World Patent Information*, *30*(1), 53–58. doi:10.1016/j.wpi.2007.05.009

Tantawi, K. H., Sokolov, A., & Tantawi, O. (2019). Advances in industrial robotics: From industry 3.0 automation to industry 4.0 collaboration. *2019 4th Technology Innovation Management and Engineering Science International Conference (TIMES-ICON)*, (pp. 1–4). IEEE.

Tao, F., Qi, Q., Wang, L., & Nee, A. Y. C. (2019). Digital twins and cyber–physical systems toward smart manufacturing and industry 4.0: Correlation and comparison. *Engineering (Beijing)*, *5*(4), 653–661. doi:10.1016/j.eng.2019.01.014

Tarallo, A., Mozzillo, R., Di Gironimo, G., & De Amicis, R. (2018). A cyber-physical system for production monitoring of manual manufacturing processes. *International Journal on Interactive Design and Manufacturing*, *12*(4), 1235–1241. doi:10.100712008-018-0493-5

Tucker, R. S. (2010). Green optical communications—Part I: Energy limitations in transport. *IEEE Journal of Selected Topics in Quantum Electronics*, *17*(2), 245–260. doi:10.1109/JSTQE.2010.2051216

Vatankhah Barenji, A., Li, Z., Wang, W. M., Huang, G. Q., & Guerra-Zubiaga, D. A. (2020). Blockchain-based ubiquitous manufacturing: A secure and reliable cyber-physical system. *International Journal of Production Research*, *58*(7), 2200–2221. doi:10.1080/00207543.2019.1680899

Wan, J., Chen, M., Xia, F., Di, L., & Zhou, K. (2013). From machine-to-machine communications towards cyber-physical systems. *Computer Science and Information Systems*, *10*(3), 1105–1128. doi:10.2298/CSIS120326018W

Wan, J., Yan, H., Liu, Q., Zhou, K., Lu, R., & Li, D. (2013). Enabling cyber–physical systems with machine–to–machine technologies. *International Journal of Ad Hoc and Ubiquitous Computing*, *13*(3–4), 187–196. doi:10.1504/IJAHUC.2013.055454

Wang, L., Liu, S., Liu, H., & Wang, X. V. (2020). Overview of human-robot collaboration in manufacturing. *Proceedings of 5th International Conference on the Industry 4.0 Model for Advanced Manufacturing: AMP 2020*, (pp. 15–58). Springer. 10.1007/978-3-030-46212-3_2

Yin, Y., Stecke, K. E., & Li, D. (2018). The evolution of production systems from Industry 2.0 through Industry 4.0. *International Journal of Production Research*, *56*(1–2), 848–861. doi:10.1080/00207543.2017.1403664

Chapter 11
Analysis of the Salaries Granted by the Maquiladora Industry and the Effect in Poverty Indicators in Tamaulipas

Olegario Mendez Cabrera
https://orcid.org/0000-0002-0126-4775
Tamaulipas Autonomous University, Mexico

Jimena Sánchez Saavedra
Tamaulipas Autonomous University, Mexico

Daniel Ávila Guzmán
https://orcid.org/0009-0008-9547-7544
Tamaulipas Autonomous University, Mexico

Abdiel Vázquez Martínez
Tamaulipas Autonomous University, Mexico

ABSTRACT

An exploratory analysis is presented about the salaries granted by the maquiladora industry in Tamaulipas and its effect on the poverty indicators, with the claim that the models of linear regression always determine the consistency in both sign and statistical validity, as well as the magnitude of the impact. Five relationships were raised through the ordinary least squares; finding that, the coefficient does not meet the expected sign, that is, it should in terms of the intuitive logic have a negative sign,

DOI: 10.4018/978-1-6684-8088-5.ch011

given that the interpretation is that the increase in salaries causes an increase in the levels of poor people, people with moderate poverty, and vulnerable by income. The only relationship that satisfies both the sign and the individual significance is that people with extreme poverty are indicative that the salary allows them to get out of that level. It is suggested that additional observations need to be collected throughout time, given the lack of these available, to be able to validate the proposed.

1. INTRODUCTION

In this document, an explorative analysis is presented about the salaries granted by the maquiladora industry in Tamaulipas and the effects on the poverty indicators indicated by the National Council for the Evaluation of Social Development Policy (CONEVAL). It is suggested that it is relevant to identify whether the greater sectoral growth, implies a decrease in the poverty indicators, considering the postulate of Campos and Monroy (2016), given the influence the maquiladora sector has in Tamaulipas and the granted salaries, with concerns that due to the fact that economic growth can reduce poverty, depending on the characteristics of the capacity of production of a region. In addition, different approaches of growth, employment, salary, welfare, and poverty, and given that Ibarra and Garcia (2016) point out that, the principal advantage of the maquiladora industry has been the creation of employment, developing diversity in studies regarding the estimation of causal models trying to explain the behavior from the employment of this sector to descriptive studies by genre. Reason for which it is proposed that the salaries granted by the maquiladora industry in Tamaulipas provide a decrease in poverty indicators.

The data from the Manufacturing, Maquiladora, and Export Service Industry (Industria Manufacturera, Maquiladora y de Servicio de Exportacion (IMMEX)) was the source used to obtain the salaries, the number of people in administrative positions and operators of seven biannual periods. For this purpose, a linear relation analysis is used through the Ordinary Least Squares method for five linear relations to adjust the salary variables (LOGTDS) compared to the different strata of poverty, such as Poor People (PP), People in Moderate Poverty (PPM), People in Extreme Poverty (PPE), Vulnerable People due to Deficiencies (PVC) and Vulnerable People due to Income (PVI).

The findings demonstrated, contrary to what was expected especially in models one (PP=f(LOGTDS)), two ((PPM=f(LOGTDS)) and five ((PVI=f(LOGTDS)), the coefficient failed to demonstrate the expected result, that is, it should in terms of the intuitive logic have a negative trend, given that the interpretation results in and increases in wages causes and increase in levels of Poor People, People with

Moderate Poverty, and Vulnerable People due to Income. The only relation that satisfies both the sign, as well as the individual significance is the People with Extreme Poverty (PPE=f(LOGTDS)), demonstrating that the salary allows them to get out of this level.

As a final suggestion, it is necessary to reinforce this document with more information, given the scarcity of information available, to be able to establish the linear relations, and that in this sense, could explain the non-relationship between the equations where the sign is not the expected one, as well as its statistical significance. In such a way, it is recommended, of course, that, over time, more observations are collected to validate what is proposed.

1.1 Problem Statement

The state of Tamaulipas is in the evolution of its main economic activity since the eighties. The entity passed to be a manufacturing region with more than 400 maquiladoras (moving to fifth place nationally) by displacing its traditional agricultural vocation. Social deficiencies have decreased constantly since the start of the present century. The principal advances are maintained in the combat of the educational deficiencies, health, living, in addition to the lack of basic drainage, water, and electricity services. According to the National Council for the Evaluation of Social Development Policy -CONEVAL (2015), the state's population in poverty went from being from 37.9% in 2014 and within the same year the population in extreme poverty decreased to 4.3%. (Government of the State of Tamaulipas, 2017, pp. 151-155).

Based on data from CONEVAL (2020), it is noted that the situation of multidimensional poverty[1], in 2020, 34.90% of the residents of the state of Tamaulipas, is in some situation of poverty 1,233.9 (thousands) people. Another important part is in a moderate poverty level[2], this is in 1,101.0 (thousands) (31.1%) people. In an extreme poverty situation[3] it is indicated that there are 135.5 (thousands) people representing 3.8%. Another data is indicative of a vulnerable population due to social deprivations[4], with 21.0%, (740.8 thousands of people). The 13.4% (represented by the 472.0 thousand of people) of the population is found in a vulnerable situation due to income[5], and 30.7%, (equals to 1,084.6 thousand of people) is not poor and not vulnerable[6]. However, in the same document, the indicators of social deprivation are indicated about the educational delay with a 15.6%, with access to health services at 16.6%, it should be noted that the highest indicator is Access to social security with 39.8%. Access to food has 14.1% of the population and with lower levels-without ceasing to be important- 5.6% la quality and space of living and finally 8.1% the access to basic services in the living space.

In the State Development Plan 2016-2022, the importance of the special programs is manifested that help a minimization of poverty indicators, indicating that the productive sectors allow this decrease, however, with data by CONEVAL (2020) themselves, no such situation is observed. Although it is clear the effect the pandemic associated with Covid-19, caused the federal government, considered activities nonessential in some sectors, bringing a drawback on the wages, because of low working hours paid and loss of employment, causing an increase in the flagged indicators. For this reason, this study is carried out to determine if the wages granted by the maquiladora industry allow them to be reduced.

1.2 Justification

The Manufacturing, Maquiladora, and Exportation Services industry, forms part of one of the most important sectors of the Municipality, and the incidence in the aspect of job creation, develops effects on the quality of life of people. Said establishments known as Maquiladoras, and according to Langle, Mendez & Sanchez (2021), the northern frontier of Mexico has been characterized by having commercial ties with the United States of North America and given the implementation of the Mexican model of import substitution, the development of economic activity in the border region has boosted.

According to INEGI and IMMEX's database (2022), in the state 364 maquiladora companies exist, employing 261,938 workers, of which 30,993 are administrative employees and 230,945 are operators. The wages granted are $1,126,952 for administrative and $2,607,417 for operators.

Given the level of economic flow, it is relevant to identify whether the greater sectoral growth, given the influence by the maquiladora sector in Tamaulipas and the granted salaries, implies a decrease in the poverty indicators.

1.3 Objectives

General

To figure out if the salary granted by the maquiladora industry to the workers in Tamaulipas fosters the environment to reduce poverty indicators.

Specifics

- Discover the labor income provided by the maquiladora industry
- Find out the poverty indicators of Tamaulipas
- Determine if economic growth has led to a decrease in the different levels of poverty.

1.4 Hypothesis

The wages given by the maquiladora industry in Tamaulipas provide a decrease in the poverty indicators.

2. THEORETICAL FRAMEWORK

Economic growth, seen as the expansion sustained by the Production Possibilities Frontier (FPP) and measured by the increase in the PIB, in a specific period, according to Parkin & Loria (2010), provides an estimated indicator of a person's standard of living known as GDP per capita, if this grows, it is presumed that people can potentially improve their well-being or reduce their levels of poverty.

As Campos & Monroy (2016) point out, considering Ravallion and Datt, 2002; Montalvo and Ravallion, 2010; Loayza and Raddatz, 2010, indicate that minimizing poverty will depend on the characteristics of the productive capacity of a region. That is, depending on the strength of its productive sectors with the highest growth, while generating jobs and their type of economic conditions, will be the level of poverty reduction.

For their part, Martínez, Flamand & Hernández (2021), determine that the economic development of a region is a topic of interest since it has a direct impact on the quality of life of its inhabitants. The relevant aspects for this development start from a broad vision, that it is necessary to incorporate aspects of the functioning of the economy, the environment, culture, and social and institutional issues, considered crucial elements to allow a better economic performance and thus raise the standard of living of the population.

The discussion of space is linked to development, and according to Friedman & Weaver (1979) and pointed out by Martínez, Flamand & Hernández (2021), they consider that the concept of space is based on the fact that production has sociopolitical, environmental, and social repercussions. Economical in the place where it occurs. Or as they indicate, according to Garofoli (1992), under the fundamentalist expression that it is simply the support of productive activity, that is, the place where economic functional forces occur and are defined, as a consequence of their logic, internal functional, location of activities, and the structure and hierarchy of the economic regions.

Therefore, the definition of the space, determined as a territory, socially characterized and determined by the entities that participate there and their relationships, allows us to identify the specialization of each region and differentiate the development options that are available there, thus, starting from this concept can be abstractly presumed to be the concept of development in dimensions, dynamics and defined variables, which facilitate the actions that promote it (Martínez, Flamand, & Hernández, 2021).

Economic growth has been one of the topics of great interest since the origins of economic science, according to Jiménez (2011), not only because of the desire for research but also because of the implications for the well-being of society. It indicates that the theory of growth analyzes the expansion of the production of economies in the long term, focusing its attention on the causes and determinants, as well as its limitations. In general, when one hears about economic growth, the concepts of economic activity, inflation, unemployment, poverty, or wages are related, among other topics, hence it is emphasized that said growth helps in most cases to reduce the number of poor or minimize the indicators of this, because the salary tends to identify people with more money, providing them with the possibility of improving their standard of living. It also points out that the analysis of growth focuses on the Gross Domestic Product (GDP) and its evolution in a given period, and from there, the per capita GDP is deduced as a better approximation (average) to the level of well-being of the population, since it's found according to Jones (2002), correlated with other indicators of quality of life.

Through economic openness, in the case of Mexico, a growth-economic policy relationship has been identified, which has allowed accelerating growth, through two axes, namely, technology and capital-labor intensity. With technological improvements, labor efficiency increased and with capital intensity, better conditions of social welfare were created. Thus, with the emergence of endogenous growth models that determined the increase in GDP per capita from within, the interest of public policies on growth was renewed, promoting the accumulation of human capital, seen as more education, access to health services, health, and nutrition, such as the promotion of capital investment and research and development (Jiménez, 2011).

Regarding the growth of the manufacturing industry and, therefore, the generation of employment, Escobar (2011), citing Sobrino (2003), determines that the concentration of economic activities in urban areas, whose explanation is given by "economies of agglomeration" and it implies a productive dynamic in a territory that is the result of decisions made by companies and individuals, seeking to maximize their benefits and profits, and which refer to internal and external advantages taken advantage of by the former and having repercussions on the latter. Referring to Mendoza (2003), Mexico has shown empirical evidence that the growth of employment in the manufacturing industry is attributed to urban wages, changes in technology, and externalities that were created by the agglomeration and specialization of the urban environment.

In addition, considering Marshall (1927), specialized environments promote industrial growth, derived from the free transfer of knowledge between employees, unlike Jacobs (1969) who establishes that externalities are the product of the economic diversity of cities and that technological progress is transmitted in an accelerated way in diversified environments. Krugman (1991) argues that this agglomeration results from the size of the market, economies of scale, and transportation costs (Escobar, 2011).

The main source of income for workers according to Gómez-Rodriguez, Rios-Bolivar & Aali-Bujari (2018) is salary, and this creates an effect on the standard of living, bringing with it a relationship between salary growth and growth of the economy, this link figures an important role in a progressive increase in inequality between countries and society and a growing need to stimulate it, likewise, indicating that wages are the "key" determinants of the living standards of the population that receives them.

Higher wages lead to greater efforts by the worker and therefore increase production in the company or industry, considering the models of Shapiro & Stiglitz (1984) and Solow (1979), indicated by Gómez-Rodriguez, Rios-Bolivar & Aali. -Bujari (2018), noting that the efficient wage is important in developing countries, in his study under the hypothesis of the efficient wage as a response to the effectiveness of stability and structural adjustment policies.

Continuing with Gómez-Rodriguez, Rios-Bolivar & Aali-Bujari (2018), they argue that the implications of the efficient wage have been extensively studied, but that few efforts have been made to empirically test the relationship between wages and the real performance of the economy with actual information. Likewise, they rescue the hypothesis of Sabia (2015) in relation to the fact that the minimum wage can be the engine of growth, demonstrating with their results that it can boost growth at the macroeconomic level if productivity is directed to sectors with better skills, probably because better training is promoted to less skilled employees.

The increases in labor productivity and wages have had a disconnect, according to Oceguera (2022), explained in various ways, with the ideas of Barkai (2020) and Ibarra and Ros (2019), pointing out that, since the capital also decreases, the stagnation of wages is due to companies having higher profit margins, thereby creating a market failure, where workers lose bargaining power, leading to lower wages.

On the one hand, according to Goss and Manning (2003) and described by García & García (2020), they establish that the quality of work, as well as the progressive relationship with technology, is defined by average salaries. On the other hand, the stagnation of real income has increased recently in the words of Oseguera (2022), and in absolute terms remunerations have remained constant, even though labor productivity has shown substantial increases, contrasting what the author says with the Kaldorian theory (Kaldor, 1961) of economic growth, where the facts establish that the payment made to the workers must be directly related to the productivity of the company.

The main advantage of the maquiladora industry has been in the creation of jobs, developing a diversity of studies in this regard, from the estimation of causal models trying to explain the behavior of employment in this sector to descriptive studies by gender. One approach has been empirical models analyzing functional relationships of interest or for the study of short-term dynamic relationships in the evolutionary process of jobs. Another group of studies has aimed to analyze the functional relationships of employment with the salary provided (Ibarra & García, 2016).

The Secretary of Economy (2010) points out that the maquiladora industry is an instrument through which it is allowed to temporarily import the necessary goods to use in an industrial or service process destined for the elaboration, transformation, or repair of merchandise of foreign origin imported temporarily for export or the provision of export services, without covering the payment of the general import tax, value-added tax (VAT), and, where appropriate, compensatory duties.

3. METHODOLOGY

The quantitative approach for Hernández (2014), is the one that uses data collection to test hypothesis based on numerical measurement and statistical analysis, to establish behavior patterns and test theories. Bernal (2010) points out that this approach is based on the measurement of the characteristics of social phenomena, which means deriving from a conceptual framework relevant to the problem analyzed, a series of postulates that express relationships between the variables studied in a deductive way. Bernal adds that this method tends to generalize and normalize results.

In relation to the above, it is evident that the approach is quantitative since information on the number of people employed by type and amount of salary received is used, as well as poverty indicators. In addition, the econometric technique is used through the method of ordinary least squares, to establish the relationship and marginal propensity of change between the indicated variables.

A series of simple or partial regressions is proposed to indicate the relationship between the exogenous variable (the logarithm of wages) and the different levels of poverty using the Ordinary Least Squares method, which is supported by a series of assumptions. According to Gujarati & Porter (2010), these are:

Assumption 1. The regression model is linear in the parameters, although it may or may not be linear in the variables, as shown in the equation $Yi= \beta1 + \beta2\, Xi + ui$

Assumption 2. Fixed values of X, or values of X independent of the error term: The values taken by the regressor X can be considered fixed in repeated samples.

Assumption 3. The mean value of the disturbance ui is equal to zero: Given the value of Xi, the mean or expected value of the random disturbance term ui is zero. Symbolically, we have that E(ui |Xi) = 0.

Assumption 4. Homoscedasticity or constant variance of ui: The variance of the error or disturbance term is the same regardless of the value of X. Symbolically, σ2.

Assumption 5. There is no autocorrelation between the disturbances: Given any two values of X, Xi, and Xj (i ≠ j), the correlation between any two ui and uj (i ≠j) is zero.

Assumption 6. The number of observations n must be greater than the number of parameters to be estimated: Successively, the number of observations n must be greater than the number of explanatory variables.

Assumption 7. The nature of the X variables: Not all X values in a given sample must be equal. Technically, var(X) must be a positive number. Furthermore, there cannot be outliers of the variable X, that is, large values in relation to the rest of the observations.

And thanks to the Gasuss-Markov theorem, these parameters (β_i) have the MELI property, that is, they are the Best Linearly Unbiased Estimators, otherwise, they have Minimum Variance. They also determine that Linear Regression deals with the study of the dependence of a variable (dependent variable) concerning one or more variables (explanatory variables) with the aim to estimate or to predict the average or population average value of the first in terms of the known or fixed values (in repeated samples) of the latter.

A large part of the econometric analysis starts from the premise that y and x are two variables that represent some population and that y is intended to be explained in terms of x, or the study of how the variations of y are explained by x, establishing an equation that relates and with x, as the following relationship:

$$y = \widehat{\beta_1} + \widehat{\beta_2} x + \widehat{u_i}$$

Where y is the dependent variable, and x is the independent variable, u is the error or disturbance term, the estimators $\left(\widehat{2}_1\right)$ and $\left(\widehat{2}_2\right)$ are identified as the intercept and the marginal proportion of change, respectively (Wooldridge, 2013).

4. RESULTS

Through the Simple Linear Regression technique and in a partial way, the data collected from the National Council for the Evaluation of Social Development

Policy (CONEVAL) used, namely, the number of people in the state of Tamaulipas in the different strata of poverty, Poor People (PP), People in Moderate Poverty (PPM), People in Extreme Poverty (PPE), Vulnerable People due to Deficiencies (PVC) and Vulnerable People due to Income (PVI). With information from the Ministry of Economy, the database of the Manufacturing, Maquiladora, and Export Service Industry (IMMEX) was considered, using both the number of workers in the industry, either in administrative or production positions, equally in this way, the total wages paid by the industries were obtained, transforming said variable into a natural logarithm (LOGTDS). All this information was obtained from seven biannual periods spanning from 2008 to 2020, which is shown in Table 1.

Table 1. Data set.[7]

Years	LOGTDS (x)	PP (y1)	PPM (y2)	PPE (y3)	PVC (y4)	PVI (y5)
2008	16.3534621	1097.70	940.55	157.16	1,073.13	254.19
2010	16.3260903	1301.70	1,118.32	183.38	899.02	312.83
2012	16.4789271	1315.60	1,155.46	160.16	918.67	303.03
2014	16.6602013	1330.70	1,179.10	151.61	855.04	393.78
2016	16.8716315	1156.20	1,052.62	103.60	1,027.0	362.87
2018	17.1538767	1287.90	1,166.08	121.81	943.41	485.36
2020	17.2182076	1233.90	1098.40	135.50	740.80	472.00

Sources: Own elaboration based on IMMEX (2022) and CONEVAL (2022)

With the obtained data, five partial regressions were carried out, taking the variable LOGTDS (Natural Logarithm of Total Salaries) as the exogenous variable and the types of poverty as the endogenous variables. Now, in Image 1, the result of the first regression is shown, considering PP= $\left(\hat{2}_1\right)-\left(\hat{2}_2\right)$ LOGTDS + $\left(\hat{u}_i\right)$ (1), thus it is expected that LOGTDS is inversely correlated with PP, given the assumed logic.

On one hand, the results indicate that, for every one-unit increase in wages, poverty increases, with the beta coefficient two having a positive sign, PP= 1058.45275+11.2293LOGTDS+ $\left(\hat{u}_i\right)$ (1), which does not agree with As expected, the t statistic value is very high, both the global test and the individual test are not statistically significant. On the other hand, Graph 1 shows the dispersion of the data, and it is evident that it does not show a trend. It is worth noting that the purpose of this analysis is exploratory, that is, over time, given the scarcity of information the authors expect to fortify them.

Figure 1. Regression poor people and wages
Source: *Own elaboration, based on IMMEX (2022) and CONEVAL (2022)*

Regression Statistics	
Multiple correlation coefficient	0.046594895
Determination coefficient R^2	0.002171084
R^2 adjusted	-0.197394699
Typical error	96.92993765
Observations	7

variance analysis	Degrees of freedom	Sum of squares	average of the squares	F	Critical value of F
Regression	1	102.2130753	102.2130753	0.01087904	0.920983891
Residuals	5	46977.06407	9395.412814		
Total	6	47079.27714			

	coefficients	Typical error	t-statistic	Probability	lower 95%	upper 95%	lower 90.0%	upper 90.0%
interception	1058.452747	1800.807606	0.587765591	0.58223101	-3570.670573	5687.57607	-2570.26169	4687.167183
Variable X 1	11.2293171	107.6609113	0.104302638	0.92098389	-265.5218659	287.9805	-205.712627	228.1712613

Figure 2. Relationship between poor people and wages in Tamaulipas 2008 – 2020
Fuente: *Own elaboration, based on IMMEX (2022) and CONEVAL (2022)*

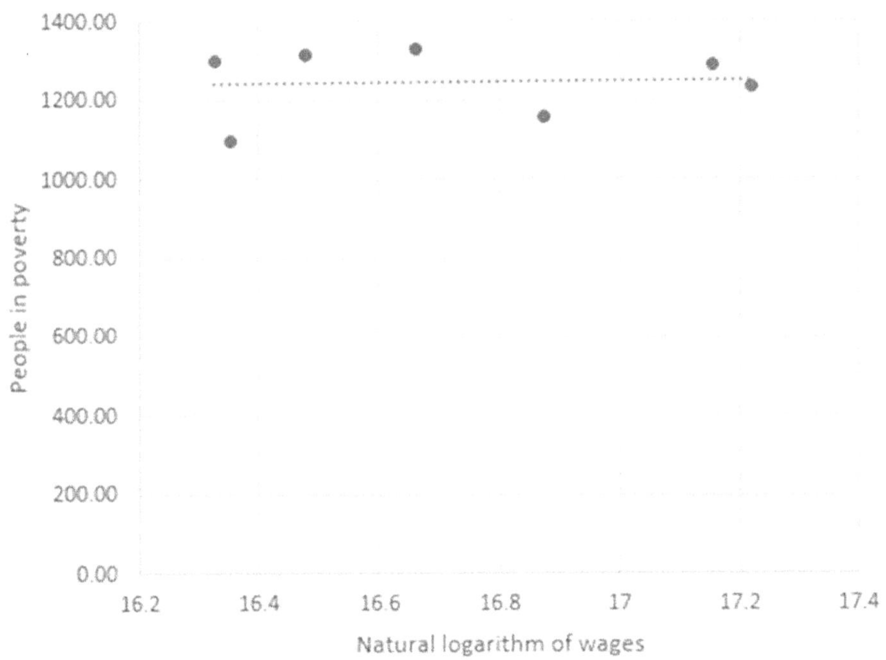

Image 2 indicates the result of the second regression, here we considered PPM= $\left(\widehat{2_1}\right)-\left(\widehat{2_2}\right)$ LOGTDS+ $\left(\widehat{u_i}\right)$ (2), a priori, it is expected that LOGTDS is inversely correlated with PPM, given the approach logical.

Figure 3. Regression people with moderate poverty (PPM) and wages
Source: *Own elaboration, based on IMMEX (2022) and CONEVAL (2022)*

Regression Statistics	
Multiple correlation coefficient	0.292566831
Determination coefficient R^2	0.085595351
R^2 adjusted	-0.097285579
Typical error	87.32077121
Observations	7

variance analysis

	Degrees of freedom	*Sum of squares*	*average of the squares*	*F*	*Critical value of F*
Regression	1	3568.75619	3568.75619	0.468038688	0.524303965
Residuals	5	38124.58542	7624.917084		
Total	6	41693.34161			

	coefficients	*Typical error*	*t-statistic*	*Probability*	*lower 95%*	*upper 95%*	*lower 95.0%*	*upper 95.0%*
interception	-8.123761922	1622.284226	-0.005007607	0.996198177	-4178.338126	4162.090602	-4178.338126	4162.090602
Variable X 1	66.35269357	96.98792789	0.684133531	0.524303965	-182.9627121	315.6680993	-182.9627121	315.6680993

It is observed that, for every one-unit increase in wages, poverty increases, with the beta coefficient having a positive sign the same as in regression (1), PPM= -8.12376+66.3526LOGTDS+$\left(\widehat{u_i}\right)$ (2), that does not agree with what was expected, the t statistic value is very high, as well as the F statistic, both of which are not statistically significant. This indicates that for every one-unit increase in wages, we see a positive increase in moderate poverty, which is unexpected.

In the same way that in Graph 1, it is observed in Graph 2, the dispersion of the data, here it can be predicted that the data behave contrary to what is expected of the salary effect on the PPM.

In relation to the approach PPE= $\left(\widehat{{}^{2}_{1}}\right)-\left(\widehat{{}^{2}_{2}}\right)$ LOGTDS+$\left(\widehat{u_i}\right)$ (3), that is, that wages have an inverse influence concerning people in extreme poverty, in Image 3, indicates the result From this regression, here it was considered, as in the previous regressions, a priori, that LOGTDS is now inversely correlated with PPE, given the approach.

It can be seen that, for every one unit increase in wages, extreme poverty decreases, the coefficient has the expected (negative) sign, PPM= 1066.7340-55.1323LOGTDS+$\left(\widehat{u_i}\right)$ (3), in this case, it is consistent with as expected, the t statistic is valid, as well as the F statistic, both at 95% confidence. This indicates wages and extreme poverty have a negative relationship, as observed in the regression data, each increase in one beta unit two decreases by -55.13. And evidently, in Graph 3, the inverse relationship between PPM and LOGTDS can be seen.

Figure 4. Relationship between people in moderate poverty and wages
Source: *Own elaboration, based on IMMEX (2022) and CONEVAL (2022)*

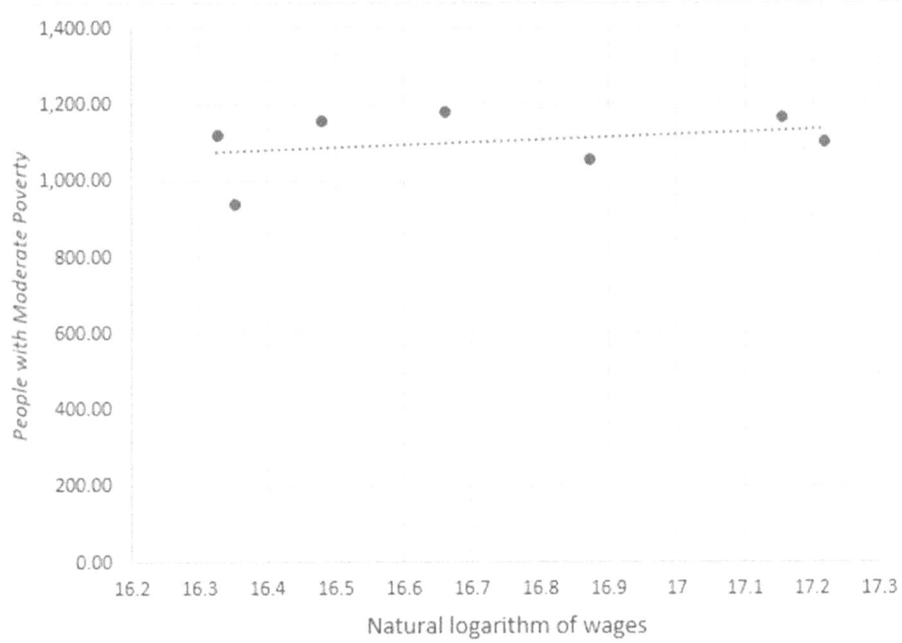

Figure 5. Regression of people in extreme poverty and wages
Source: *Own elaboration, based on IMMEX (2022) and CONEVAL (2022)*

Regression Statistics	
Multiple correlation coefficient	0.762690875
Determination coefficient R^2	0.58169737
R^2 adjusted	0.498036844
Typical error	18.82427302
Observations	7

variance analysis

	Degrees of freedom	*Sum of squares*	*average of the squares*	*F*	*Critical value of F*
Regression	1	2463.842463	2463.842463	6.953068528	0.046149919
Residuals	5	1771.766273	354.3532546		
Total	6	4235.608736			

	coefficients	*Typical error*	*t-statistic*	*Probability*	*lower 95%*	*upper 95%*	*lower 95.0%*	*upper 95.0%*
interception	1066.734034	349.7257384	3.050201677	0.028417775	167.7354031	1965.73266	167.7354031	1965.732664
Variable X 1	-55.13236035	20.90828114	-2.636867181	0.046149919	-108.8788081	-1.38591263	-108.8788081	-1.385912629

In the same sense, the approach PVC= $\left(\widehat{{}^{2}_{1}}\right)-\left(\widehat{{}^{2}_{2}}\right)$ LOGTDS+ $\left(\widehat{u_i}\right)$ (4), a negative sign of beta 2 is expected a *priori*, that is, that wages have an inverse influence concerning people vulnerable due to deficiencies, Image 4 shows the result of this regression.

Figure 6. Relationship between people in extreme poverty and wages
Fuente: *Own elaboration, based on IMMEX (2022) and CONEVAL (2022)*

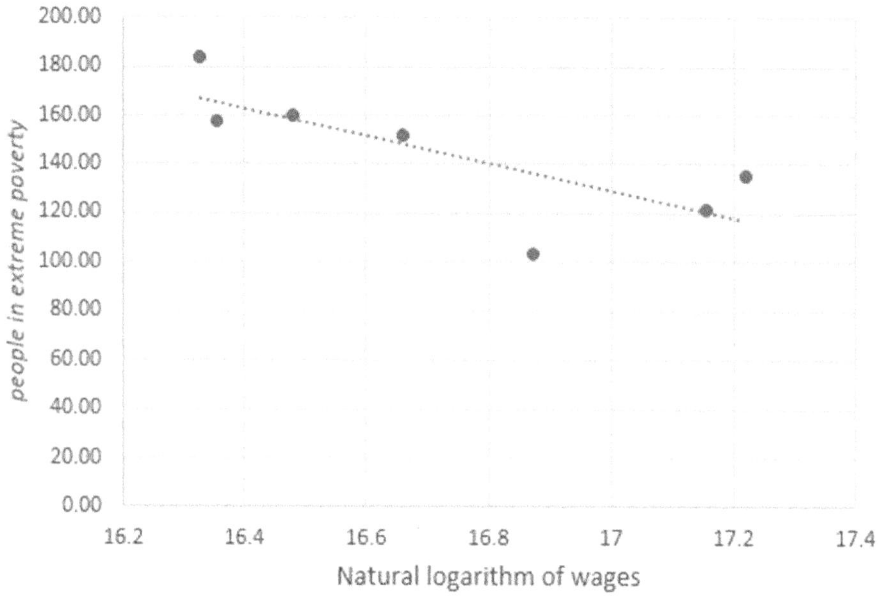

Figure 7. Regression of vulnerable people due to deficiencies and salaries
Source: *Own elaboration, based on IMMEX (2022) and CONEVAL (2022)*

Regression Statistics	
Multiple correlation coefficient	0.44073055
Determination coefficient R^2	0.194243418
R^2 adjusted	0.033092101
Typical error	107.8538006
Observations	7

variance analysis

	Degrees of freedom	*Sum of squares*	*average of the squares*	*F*	*Critical value of F*
Regresión	1	14021.1411	14021.1411	1.205348003	0.322290517
Residuos	5	58162.21151	11632.4423		
Total	6	72183.35261			

	coefficients	*Typical error*	*t-statistic*	*Probability*	*lower 95%*	*upper 95%*	*lower 95.0%*	*upper 95.0%*
interception	3121.876762	2003.756002	1.558012432	0.179966857	-2028.942021	8272.69554	-2028.94202	8272.69554
Variable X 1	-131.5199994	119.7941394	-1.09788342	0.322290517	-439.460638	176.420639	-439.460638	176.420639

The result shown in Image 4 shows that the sign is correct for the beta 2 coefficient, PVC= 3121.8757-131.5199LOGTDS+$\left(\widehat{u_i}\right)$ (4), however, the coefficient is not significant. Graph 4 indicates an inverse relationship between PVC and LOGTDS.

Figure 8. Relationship between people vulnerable due to deprivation and wages
Source: *Own elaboration, based on IMMEX (2022) and CONEVAL (2022)*

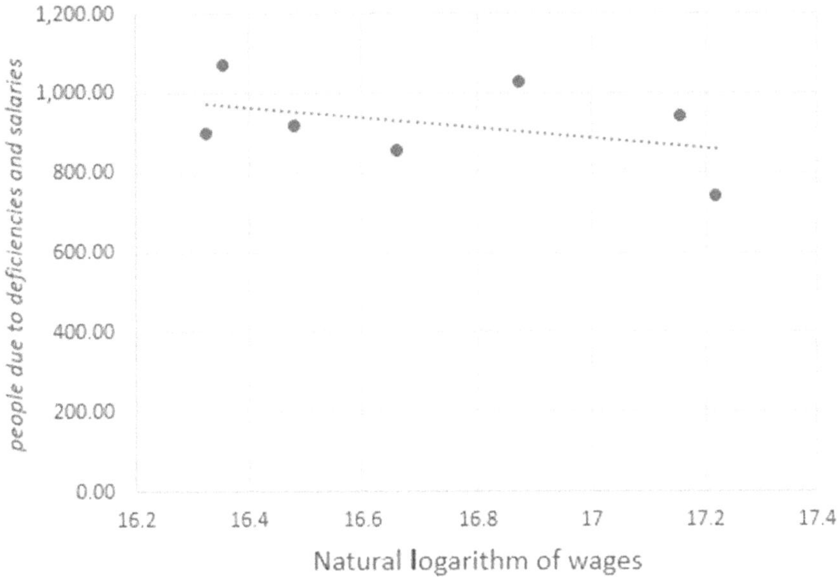

Finally, in the partial regression of the Vulnerable Population by Income versus Salary, PVI= $\left(\widehat{2}_1\right)-\left(\widehat{2}_2\right)$ LOGTDS+$\left(\widehat{u_i}\right)$ (5), can be seen in image 5.

It is observed that the beta 2 coefficient does not have the expected sign despite being significant PVI= -3334.0498+221.4409LOGTDS+$\left(\widehat{u_i}\right)$ (5), showing a positive relationship, contrary to what was expected. Graph 5 reveals this trend.

Figure 9. Regression of vulnerable people by income and wages
Source: *Own elaboration, based on IMMEX (2022) and CONEVAL (2022)*

Regression Statistics	
Multiple correlation coefficient	0.934782853
Determination coefficient R^2	0.873818982
R^2 adjusted	0.848582779
Typical error	33.8812168
Observations	7

variance analysis	*Degrees of freedom*	*Sum of squares*	*average of the squares*	*F*	*Critical value of F*
Regression	1	39748.01563	39748.01563	34.6256117	0.002013788
Residuals	5	5739.684258	1147.936852		
Total	6	45487.69988			

	coefficients	*Typical error*	*t-statistic*	*Probability*	*lower 95%*	*upper 95%*	*lower 95.0%*	*upper 95.0%*
interception	-3334.049835	629.4603542	-5.296679628	0.00320077	-4952.129188	-1715.97048	-4952.12919	-1715.97048
Variable X 1	221.4409032	37.63215746	5.884353121	0.00201379	124.7043628	318.177444	124.704363	318.177444

Figure 10. Relationship between vulnerable people by income and wages
Source: *Own elaboration, based on IMMEX (2022) and CONEVAL (2022).*

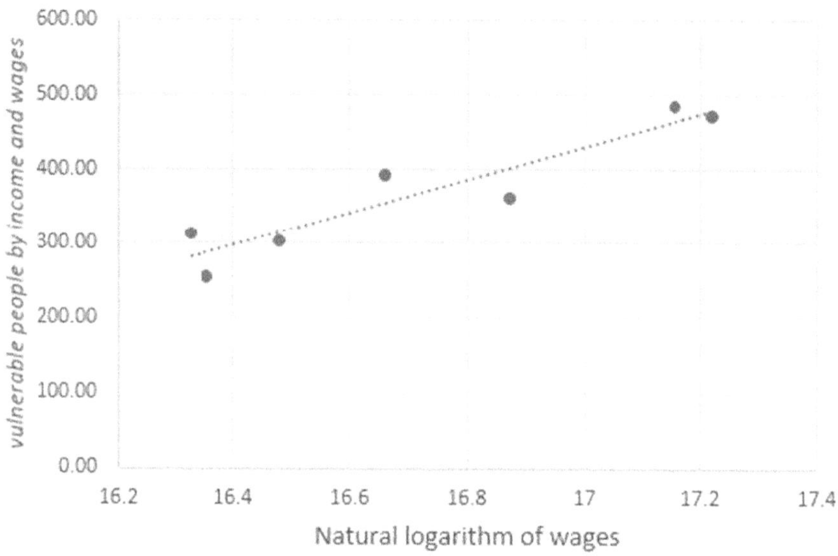

Table 2 summarizes the results of the regressions indicating the endogenous variables, Poor People (PP), People in Moderate Poverty (PPM), People in Extreme Poverty (PPE), Vulnerable People due to Deficiencies (PVC), and Vulnerable People due to Income (PVI). And as an exogenous variable, the logarithm of wages (LOGTDS).

Table 2. Statistical summary

Variable	Coefficient	Standard error	Significance (0.05)
PP	11.2293	107.6609	0.92099
PPM	66.3526	96.9879	0.52430
PPE	-55.1323	20.9082	0.04614
PVC	-131.5199	119.7941	0.32229
PVI	221.4409	37.6321	0.00201

Fuente: Own elaboration, based on IMMEX (2022) and CONEVAL (2022).

5. CONCLUSION

The general objective of this work is to find out if the salary granted by the maquiladora industry to workers in Tamaulipas favors the environment to reduce poverty indicators. Using the Ordinary Least Squares Method, five linear regression models were proposed, to observe if the endogenous variables responded significantly to the variations of the exogenous variable.

For this, the proposed hypothesis refers to the fact that the salaries granted by the maquiladora industry in Tamaulipas favor a decrease in poverty indicators. Starting from a theoretical context of economic growth and its main indicator (GDP) that can enhance the well-being of the population and therefore reduce poverty levels (Parkin & Loría, 2010). The productive sectors with the highest growth generate jobs and their conditions, which will reduce poverty (Campos & Monroy, 2016) and ending with the analysis of the statistical significance of the proposed models.

Table 2 indicates for the approach of the model one PP= 1058.45275+11.2293LOGTDS+$\left(\widehat{u_i}\right)$ (1), which, for every one-unit increase in wages, poverty increases, which does not agree with what was expected.

In equation two, it can be seen that, for every one unit increase in wages, Moderate poverty increases PPM=-8.12376+66.3526LOGTDS+$\left(\widehat{u_i}\right)$, the same as in equation one.

Regarding equation three, it can be seen that, for each increase in wages by one unit, extreme poverty decreases, the coefficient has the expected (negative) sign, PPM= 1066.7340-55.1323LOGTDS+$\left(\widehat{u_i}\right)$ (3), In this case, it is as expected, the t statistic is valid, as well as the F statistic, both at 95% confidence. This indicates that wages and extreme poverty have a negative relationship, as observed in the regression data, for each increase in one beta unit, two decreases by -55.13.

In model four, it can be seen that the sign is correct for the beta 2 coefficient, PVC= 3121.8757-131.5199LOGTDS+$\left(\widehat{u_i}\right)$ (4), however, the coefficient is not significant.

The approach of model five, the beta 2 coefficient, does not have the expected sign despite being significant PVI= -3334.0498+221.4409LOGTDS+$\left(\widehat{u_i}\right)$ (5), showing a positive relationship, contrary to what was expected.

In relation to the results, contrary to what was expected, especially in models one, two, and five, the coefficient does not meet the expected sign, that is, it should, in terms of intuitive logic, have a negative sign, given that the interpretation. It will be that the increases in wages cause increases in the levels of Poor People, People with Moderate Poverty and Vulnerable People by Income.

The only relationship that satisfies both the sign and the individual significance is that of People in Extreme Poverty, indicative that the Salary allows them to get out of this level, complying with..." guarantee of the exercise of three or more of their rights for the social development (has three or more deficiencies, out of six possible, within the Social Deprivation Index) and, in addition, is below the line of extreme poverty by income. The latter means that if you used all the household's disposable income, you would not be able to purchase the food basket (CONEVAL, 2020)".

In addition, as already stated at the beginning, another of the factors that can affect these relationships and especially the inhabitants, was the pandemic effect associated with Covid-19, due to non-essential activities, few paid working hours, loss of employment and/or the reallocation of salary to cover the health recommendations in the purchase of gel, face masks and medicines.

It is important to emphasize that this document is exploratory in nature, that is, that, based on the available information, it is the closest to establishing linear relationships and that in this sense, the non-relationship between the equations where the sign is not could be explained. expected, as well as its statistical significance. In such a way, it is recommended, of course, that, over time, more observations are collected in order to validate what is proposed.

REFERENCES

Campos, R., & Monroy, L. (2016). La relación entre crecimiento económico y pobreza en México. *Investigacion Economica*, *75*(298), 77–113. doi:10.1016/j.inveco.2016.11.003

Escobar, A. (2011). Determinantes del empleo en la industria manufacturerar en México. *Papeles de Población*, 251–276.

García, G., & García, G. (2020). Polarización salarial y ocupacional en México. *Estudios Económicos*, 153-188.

Gobierno del Estado de Tamaulipas. (2017). *Plan Estatal de Desarrollo 2016-2022*. Victoria, Tamaulipas, México.: Gobierno del Estado de Tamaulipas.

Gómez-Rodríguez, T., Ríos-Bolivar, H., & Aali-Bujari, A. (2018). Salario eficiente y crecimiento económico para el caso de América LAtina. *Ensayos*, 213-235.

Gujarati, N., & Porter, D. (2010). *Econometría*. McGraw-Hill.

Ibarra, J., & García, F. (2016). Las demandas de factores productivos en la industria maquiladora. *Estudios Económicos*, 265-303.

Instituto Nacional de Estadística y Geografía. (2015). *Síntesis metodológica de la estadística del programa de la industria.* INEGI. https://www.inegi.org.mx/est/contenidos/proyectos/accesomicrodatos/immex/default.aspxInstituto Nacional de Geografía y Estadística. (2022). *Manufacturas.* INEGI. https://www.inegi.org.mx/temas/manufacturas/

Jiménez, F. (2011). *Crecimiento Económico: Enfoques y Modelos.* Lima, Perú: Fondo editorial de la Pontificia Universidad Católica del PErú.

Langle, M., Méndez, O., & Sánchez, J. (2021). Factores predictores del índice de rotación de personal: el caso de una empresa maquiladore en Reynosa. *Análisis Económico*, 119-140.

Loayza, N. V., & Raddatz, C. (2010). The composition of growth matters for poverty alleviation. *Journal of Development Economics, 93*(1), 137–151. doi:10.1016/j.jdeveco.2009.03.008

Martínez, S., Flamand, L., & Hernández, A. (2021). Creimiento y desarrollo en los municipios frotnerizos del norte de México: Índice de Desarrollo Municipal Básico y PIB. In *COLEF, 25 años de integración económica en la frontera norte de México (pág. 411).* Colegio de la Frontera Norte.

Montalvo, J. G., & Ravallion, M. (2010). The pattern of growth and poverty reduction in China. *Journal of Comparative Economics, 38*(1), 2–16. doi:10.1016/j.jce.2009.10.005

Oseguera, A. (2022). Concentración en el mercado laboral y su relación con los salarios en México. *Estudios Económicos*, 45-102.

Parkin, M., & Loría, E. (2010). *Microeconomía. Versión para Latinoamérica.* PEARSON EDUCACIÓN.

Ravallion, M., & Datt, G. (2002). Why has economic growth been more pro-poor in some states of India than others? *Journal of Development Economics, 68*(2), 381–400. doi:10.1016/S0304-3878(02)00018-4

Reynosa, M. (2021). *Plan Municipal de Desarrollo 2021-2024.* Municipo de Reynosa Tamaulipas.

Secretaria de Economía. (2010). *Secretaria de Economía.* Economia. http://www.2006-2012.economia.gob.mx/comunidad-negocios/industria-y-comercio/instrumentos-de-comercio-exterior/immex

Wooldridge, J. (2013). *Introducción a la Econometría.* Cengage Learning.

ENDNOTES

1 A person is in a situation of multidimensional poverty when the exercise of at least one of their rights for social development is not guaranteed, and their income is insufficient to acquire the goods and services required to satisfy their needs.

2 It is that person who, being poor, is not extremely poor. The incidence of moderate poverty is obtained by calculating the difference between the incidence of the population in poverty minus that of the population in extreme poverty.

3 A person is in a situation of extreme poverty when the exercise of three or more of their rights for social development is not guaranteed (they have three or more deficiencies, out of six possible, within the Social Deprivation Index) and, in addition, they are below the extreme income poverty line. The latter means that if you used all of the household's disposable income, you would not be able to purchase the food basket.

4 Population that presents one or more social deprivations and whose income is greater than or equal to the well-being line.

5 That population that does not present social deprivations but whose income is less than or equal to the well-being line.

6 Population that does not present social deprivations and whose income is greater than or equal to the well-being line.

7 Las cifras están en miles, a excepción de la columna LOGTDS(x)

Chapter 12
The Salary and Wage Inequality Effect on Productivity on the Mexico-US Border:
Mexican Middle Management Supervisor Perspective

Miguel Sahagun
https://orcid.org/0000-0002-6962-4714
High Point University, USA

Fernando Ortiz-Rodriguez
https://orcid.org/0000-0003-2084-3462
Tamaulipas Autonomous University, Mexico

Jose-Melchor Medina-Quintero
Tamaulipas Autonomous University, Mexico

ABSTRACT

In today's global economy, emerging economies play a very important role in the production chain. Countries such as China and India have become competitive participants in the global market. The labor force in these emerging markets has developed new skills, leading developed economies such as the United States to move important areas of their production chain into these countries. This phenomenon had reached Mexico as well and multiple cases of salary and wage inequality could be found. Thus, this research tries to establish a frame that could help better understand the effect that salary and wage inequality play in the leadership-productivity relation among middle management supervisors within the manufacturing industry at the

DOI: 10.4018/978-1-6684-8088-5.ch012

Mexico-US border. A questionnaire was developed and applied to middle management supervisors on both sides of the border (Mexico and the US). The main contribution of this study is to help better understand the impact of salary and wage inequality among middle management supervisors.

1. INTRODUCTION

Now a days the emerging economies are key players in Global business, countries like China and India have become large and competitive participants in the global market. As a result, the labor force on these emerging markets has developed high skills, therefore multiple companies from developed economies such as the US and Germany have moved their manufacturing facilities to these countries. This phenomenon has created wage inequality between manufacturing workers from the developed economies and manufacturing workers in emerging markets. This phenomenon has reached Mexico as well.

In this sense, analyzing the effects of wage inequality in the Mexican-US border will help better understand this phenomenon and improve path of productive development among borders. In the study of Sanchez (2013) he considers that: "The theorization of the border between Mexico and the United States helps to recognize its dynamics and its hierarchical and functional attributes." This view also coincides with Nicol and Minghi's (2005) emphasis on flexible approaches capable to consider their dynamic nature when studying border regions. Theorizing the practices that are generated in the border territory between Mexico and the United States involves investigating the power behind those practices, the scales in which they operate, the ways in which the border adapts to those processes, and the discourses that result from them.

Additional studies also consider that the promotion of the borders between Mexico and the United States is fostered by international free trade agreements (Carbajal, 2017). For instance, the implementation of the North American Free Trade Agreement (NAFTA) since 1994 have furthered the dynamics of the economic activity of the northern border, especially in the automotive industry. For multinational companies in the automotive sector, NAFTA represented the possibility of a larger potential market with better commercial conditions. Additionally, Mexican Border States had become natural areas for new foreign investment due to their geographic location advantages derived from their proximity to the US market. Sánchez and Campos (2010) explained that the Mexican border region is, economically speaking, one of the most outstanding in the country. The region's proximity to the United States places it in an outstanding position to strengthen the industrialization of the area. The development of the Mexican northern border, helped by the maquiladora industry, are linked to the cycles of US manufacturing and the Mexican labor, which is much lower (López, 2004).

Underlying the effect of leadership in personal productivity among middle management supervisors there are additional factors that need to be identified and analyzed. For instance, perceived salary and wage inequality influences negative attitudes that may undermine personal productivity among workforce. Thus, salary and wage inequality can reduce or even eliminate the positive effect of leadership on personal productivity. The question salary and wage inequality raises is: how influential is salary and wage inequality on the leadership-productivity relation among middle management supervisors? Furthermore, is there an administrative process companies can implement to minimize or, if possible, eliminate the negative attitudes that undermine personal productivity among middle management supervisors? Thus, there is a need to first, determine the effect of leadership in middle management supervisors' personal productivity. Second, determine the effect of salary and wage inequality in the leadership-productivity relation among middle management supervisors. And third, determine if implementing an internal administrative process to develop and control leadership execution could overcome the negative impact of salary and wage inequality on the leadership-productivity relation.

The following research questions guide this study:

Q1: What is the effect of leadership in middle management supervisors' personal productivity?

Q2: What is the effect of salary and wage inequality in the leadership-productivity relation among middle management supervisors? And

Q3: Does implementing an internal administrative process to develop and control leadership execution overcome the negative impact of salary and wage inequality on the leadership-productivity relation?

In the following sections, the conceptual model is discussed. The theoretical underpinnings are then presented and the hypotheses formulated. In the section following, the methodology is described, which includes the research design, measurement, and sample characteristics, after which the results are rendered and discussed. Finally, theoretical and practical implications are asserted and some suggestions for further research are offered.

2. CONCEPTUAL FRAMEWORK

The globalization facilitates business activities, eliminates existing barriers and tariffs, and is a great opportunity for business growth. However, human performance is the most important element that companies should encourage (Al-Matari 2014). The human factor triggers organization's growth. Furthermore, companies had determined corporate goals through different performance measures related to a process called technical efficiency (Sheu & Yang, 2005; Bozec & Dia, 2007; Destefanis & Sena, 2007; Garcia-Sanchez, 2010).

2.1 The Role of Leadership on Personal Productivity

The theory of collective organizational commitment, integrates the resource management model (Murray et al., 2015) and analyzes the organizational collective leadership level through constructs influenced by motivation. These organizational practice is highly influenced by the transformational leadership of the CEO.

Resource management builds new capabilities and processes that maintain and create higher value for stakeholders (Sirmon et al., 2007). Throughout the use of the resource management model companies can increase the group leadership by increasing the level of involvement and work participation among members, as well as improving their relationships. Moreover, in order to further increase leadership levels in an organization, companies should treat leadership as a collective construct (Murray et al., 2015). In other words, productivity is directly related to the leadership levels of the individual.

Previous studies on incremental productivity have founded that it must be analyzed considering different human capabilities such as knowledge, skills, motivations, and internal elements that mold the employees' character and help them face day to day work related situations (Ziapour, Khatony and Jafary, 2015, López-Vázquez and Sahagún, 2016). Furthermore, the study of incremental productivity based on labor competencies is key to find new ways to increase productivity levels among companies (López-Vázquez and Sahagún, 2016). Companies' labor force commonly referred as human capital could become a sustainable competitive advantage for any company if it is managed correctly (Moreno, 2007). However, although human capital is essential to accomplish organizational objectives, it is required to have a perfect alignment between organizational objectives and human capital objectives (Moreno, 2007). Moreover, it is necessary to determine the key set of work elements in addition to human capabilities such as knowledge, skills, motivations, and internal elements that mold the employees' character that help human capital achieve companies' productivity goals and positively impact employees' performance (López-Vázquez and Sahagún, 2016). López-Vázquez and Sahagún, (2016) suggest leadership measured by recognition, autonomy, group cohesion, communication, pressure, and support as the key set of work elements that allow employees to improve their ability to overcome challenges and improve their performance on daily basis.

Consequently, leadership measured by 1) recognition, 2) autonomy, 3) group cohesion, 4) communication, 5) pressure, and 6) support is expected to have a direct effect on individual productivity among middle management supervisors (López-Vázquez and Sahagún, 2016).

Therefore, it is hypothesized:

H1: *There is a direct and positive effect of leadership measured by 1) recognition, 2) autonomy, 3) group cohesion, 4) communication, 5) pressure, and 6) support on individual productivity among middle management supervisors.*

2.2 The Effect of Salary and Wage Inequality on the Leadership-Productivity Relation

In recent years companies have had the need to make changes in their work structures due to process-oriented manufacturing, technological changes, and implementation of more efficient work systems. However, it is important to acknowledge the overall influence on work commitment among employees transmitted via transformational leadership (Raja and Palanichamy, 2011). Furthermore, Raja and Palanichamy (2011) found that implementing a transformational leadership style allows for a better organizational commitment among employees and their level of commitment does not get affected by salary differences.

On the other hand, Williams (1992) and Olsen and Richardson (1999) have expressed concerns regarding inequality implications when including productivity costs in economic evaluations. One issue is that the business development programs assumed that workers, in exchange for their salaries only, produce substantial gains for their companies due to productivity improvements. On the other hand, studies have shown that productivity costs have not been included in economic evaluations and the methods used to estimates productivity costs among companies (Krol, Brouwer, and Rutten, 2013). Thus, companies should first develop common standards with respect to the identification, measurement, and valuation of productivity costs.

Buttner and Lowe (2017) examined the different effects of the perception of equitable payment and productivity with respect to internal and external equity of payment and they discovered that the highly productive participants who perceived the equity of payment reported the greatest commitment. When the salary was inequitable, the most productive participants reported the lowest commitment. In other words, the higher the salary, the higher productivity. However, this approach is not determinant to accurately establish the effects of wage inequality and productivity.

Thus, the extent to which workers perceive salary and wage inequality has a negative effect on individual productivity among workers. Therefore, it is hypothesized:

H2: *Salary and wage inequality has a negative and significant moderating effect on the leadership-productivity relation among middle management supervisors.*

2.3 The Importance of Having an Internal Administrative Process to Control Leadership Execution Among Middle Management Supervisors

The leadership types in business have been studied under the conditions that there is a production manager and/or supervisor who is responsible for the performance results of the group and/or department he/she is in charge. Some studies had reported average performance levels among groups or departments in which members only focus on performing their jobs (Rappe & Zwick, 2006). When companies try to make employees assume their leadership roles which implies to have competence on active communication, delegation, feedback, interact with subordinates, and act as a representative of the unit, they realize it is practically impossible to achieve. Therefore, it is necessary to train the workforce to execute their leadership roles through a set of competences that belong to a clear frame of reference in order to successfully manage them. A competence is any underlying characteristic of an employee that has a retribution in cash or superior performance in his/her job (McClelland and Boyatzis, 1982, Spencer & Spencer, 1993).

On the other hand, leadership has also been studied as an active element of the organizations, visible in the organizational culture, work environment, and development and value creation of the organization. Leadership style is a variable that affects, positively or negatively, the organization's learning processes (Dzinic, 2015). The bases of organizational learning are at the individual level and individual learning has been defined as an individual competence that provides the ability and motivation to carry out the required tasks (Bontis et al., 2002). Moreover, learning is a strategic resource for the group, team, department, or organization considering that transformational leadership aimed at achieving tasks and goals toward the increase of productivity exist. In this sense, some studies have focused on the importance of leadership at the team level. The leadership style at the team level is influenced by the interactions between the leader and his/her followers (Busic, Robinson & Ramburuth, 2010).

Thus, it is necessary to manage leadership at individual and group levels in order to facilitate organizational learning and develop a participatory and transformational administration. Therefore, it is hypothesized:

H3: *Having an internal administrative process that helps develop and control the leadership execution among middle management supervisors reduces the negative impact of salary and wage inequality on the leadership-productivity relation.*

Next the research methodology is presented.

3. METHODOLOGY

The city of Reynosa was selected because it has over 160 manufacturing facilities part of multinational companies and employing more than 1,500 middle management supervisors who are aware of the wage differences between Mexico and the U.S.

3.1. Research Design

A self-administered questionnaire of 74 items and questions was employed to capture data from middle management supervisors in their natural working environment. Majority of the constructs were measured using multiple-items from existing literature (reflective measurement) and using a seven-point Likert scale anchored between "strongly disagree" (1) and "strongly agree" (7). However, the wording in the items was slightly modified to fit research purposes. The construct of wage compensation was measured using a formative measurement system.

This study relied on self-reported measures rather than direct observation. Self-reported measures have shown to be highly consistent with objective measures on many settings (Chen, Gillenson, & Sherrell, 2002).

The questionnaire was administered in Spanish for all participants because Spanish is the participants' native language. Furthermore, all manufacturing facilities used for this study were located in the Mexican side of the border. However, since majority of the items were originally written in English, a double back translation procedure was utilized to develop the Spanish version and avoid erratic literal English-language translation (Werner & Campbell, 1970).

3.2. Sample Demographics

For hypotheses testing, 22 participants were selected using non-random sampling. All the participants were middle management level supervisors from 5 manufacturing companies and part of the extractive industry; all these companies were in the Mexico-U.S. border (Mexican side). Seventy three percent of the participants were men and 27% women. Sixty-eight percent of the participants were married, 28%

single, and 4% divorced. All participants have attended and finished college and 72% have or are pursuing a master's degree. Finally, all participants are between 28 and 44 years of age.

3.3. Plan of Analysis

For the purposes of refining the instrument used in this study, the questionnaire was verified using Exploratory Factor Analysis (EFA) with Maximum Likelihood to assure measurement model reliability and validity prior to hypotheses testing. Furthermore, this process reduces the risk of utilizing inappropriate measures during hypotheses testing and misleading results. After the instrument was refined, hypotheses testing was performed using linear regression analysis and hierarchical multiple regression analysis to examine the moderation effect of wage inequality in the leadership-productivity relation among middle management level supervisors in the manufacturing industry. SPSS version 22.0 was used to perform such analysis.

3.4. Measurement Model Reliability and Validity

Cronbach's alpha values for all constructs that were captured using reflective measures (individual productivity, work autonomy, group cohesion, group communication, group pressure, group support, and group recognition) ranged from .728 to .881, and Cronbach's alpha value for the only construct that was captured using formative measures (wage compensation) was .653. However, in order to achieve these reliability values, eleven out of the sixty-four items corresponding to five constructs were removed from the initial measurement model. Items' removal was based on both statistical results and conceptual considerations.

Additionally, all VIF values between individual productivity (DV) and each independent variable (work autonomy, group cohesion, group communication, group pressure, group support, group recognition, and wage compensation) exhibit satisfactory levels (Pan & Jackson, 2008; Rogerson, 2001) ranging from 1.298 to 4.023. These results suggest lack of presence or inexistence of collinearity and multicollinearity between individual productivity (DV) and work autonomy, group cohesion, group communication, group pressure, group support, group recognition, and wage compensation (IVs).

Convergent validity was achieved by exhibiting 1) appropriate reliability among constructs measured by their Cronbach's alpha (.653 for a formative construct and .728 to .881 for reflective constructs), 2) standardized loading estimates among items equal or higher than .50 (.450 to .913); only two measurement items among 53 loaded slightly below threshold value (.450 and .486 respectively), and 3) construct average variance explained equal or higher than .50 (.405 to .666); only one construct (individual productivity) exhibited an average variance explained slightly below threshold value (.405) (Hair et al., 2010).

Finally, constructs exhibit discriminant validity by having larger average variance explain estimates than the corresponding squared inter-construct correlation estimates (Fornell & Larcker, 1981; Hair et al., 2010) or by having larger square roots of construct's average variance explained estimates than the corresponding inter-construct correlation estimates (see Table 1).

Table 1. Correlation matrix (N = 22) productivity, leadership, and wage compensation

Construct	1	2	3	4	5	6	7	8
1. Individual Productivity	**.803** [a]/ ***.637*** [b]							
2. Work Autonomy	.572**	**.801** [a]/ ***.769*** [b]						
3. Group Cohesion	.259	.247	**.881** [a]/ ***.798*** [b]					
4. Group Communication	.221	.145	.673**	**.878** [a]/ ***.780*** [b]				
5. Group Pressure	.126	-.039	-.316	-.055	**.786** [a]/ ***.782*** [b]			
6. Group Support	.315	.528*	.458*	.097	-.093	**.877** [a]/ ***.827*** [b]		
7. Group Recognition	.171	.371	.436*	.245	-.222	.769**	**.876** [a]/ ***.824*** [b]	
8. Wage Compensation	.393	.325	.336	.187	-.229	.136	.087	**.653** [a]/ ***.830*** [b]

*p<.05, **p<.01 (2-tailed).

[a]Cronbach's Alpha

[b]$\sqrt{AVE}$ (square root of average variance explained)

4. RESULTS

4.1. Leadership-Productivity Relation

Table 2 exhibits three hierarchical multiple regression models employed to test for wage compensation moderation effects. The first model (see Model 1[a] in Table 2) shows that the six factors used to measure leadership (work autonomy, group cohesion, group communication, group pressure, group support, and group recognition) explain a high squared multiple correlation coefficient for individual productivity (R^2 = .389). Thus, hypothesis H1 is supported.

4.2. Moderating Effect of Wage Compensation in the Leadership-Productivity Relation

The second model (see Model 2[b] in Table 2) adds the proposed moderating variable (wage compensation) into the previous model, which resulted in a more comprehensive explanation for individual productivity. The squared multiple correlation coefficient increased from 38.9% to 43.7%. The increment on individual productivity explanation for this model revealed a difference of 4.8%.

The third model (see Model 3[c] in Table 2) adds the interaction terms for the moderating variable introduced in the Model 2[b], and it also resulted in an a more comprehensive explanation for individual productivity. The squared multiple correlation coefficient increased from 43.7% to 60.9%. The increment on individual productivity explanation for this model revealed a difference of 17.2%. Thus, hypothesis H2 is supported.

Moreover, if the moderating variable (wage compensation) is removed from Model 3[c] and only the interaction terms and the leadership factors are kept, the square multiple correlation coefficient for individual productivity is 60.8%. In other words, the interaction terms alone provide an additional 21.9% explanation increment to Model 1[a]. These results indicate that majority of the effect of wage compensation on the individual productivity variable is via its interactions with the six factors used to capture leadership and not by itself.

4.3. Having an Internal Administrative Process to Control Leadership Execution Among Middle Management Supervisors

Two managers from different manufacturing facilities were interviewed to capture their views toward the importance of having an internal administrative process that controls leadership execution among middle management supervisors. Their responses were aligned with previous research. To achieve success in organizational performance, management must work together with their teams giving them the necessary autonomy to achieve the proposed goals and thus achieve greater work commitment by the group/team members (Appelbaum, St-Pierre, & Glavas, 1998).

Manager A: *Leadership is a basic tool, necessary for the achievement of company's objectives, it is the most important factor to achieve or fulfill productivity metrics... There is a direct correlation; the better your role as a leader, the easier it is to fulfill the productivity metrics. In other words, the greater the leadership, the greater the productivity achieved.*

Table 2. Regression results: Leadership-productivity relation and wage compensation moderation models (N = 22)

	MODEL 1[a]		MODEL 2[b]		MODEL 3[c]	
Dependent Variable: Purchase Intention	**B**	**t-value**	**B**	**t-value**	**b**	**t-value**
Constant	4.742	.406	2.441	.207	18.643	.169
Group Autonomy	.544**	2.248	.455*	1.793	2.090	.755
Group Cohesion	.166	.463	.073	.198	2.632	1.000
Communication	.065	.204	.086	.271	-3.367	-1.344
Pressure	.181	.788	.216	.935	-1.491	-.721
Support	.057	.141	.094	.233	-.569	-.152
Recognition	-.123	-.357	-.097	-.282	-.483	-.126
Wage Compensation			.249	1.093	-.524	-.112
Group Autonomy x Wage Compensation					-4.239	-.647
Group Cohesion x Wage Compensation					-5.419	-1.151
Communication x Wage Compensation					7.407	1.533
Pressure x Wage Compensation					1.995	.941
Support x Wage Compensation					1.636	.287
Recognition x Wage Compensation					-.235	-.040
R^2	**.389**		**.437**		**.609**	
F	**1.591**		**1.552**		**.959**	
ΔR^2			**.048**		**.172**	

[a]Core variable effects
[b]Moderating variable effects
[c]Two-way interaction effects
*p<.10, **p<.05, ***p<.01 (one-tailed test for hypothesized relationships).

Having individual leadership becomes the main factor to change processes; however, techniques that adequately model and reward the behavior of individuals should be used. The organization's ability to individually motivate employees is a higher level of control closely related to the reward system (Appelbaum, St-Pierre, & Glavas, 1998).

Manager A: *Work recognition is required to motivate employees. However the person requires to be fully commitment to his/her position. Furthermore, he/she needs to be passionate about his/her job and daily activities. Overall, these are the essential elements to be productive…*

Manager B: *People should be given a "well done" pat on the back as a daily common practice as well as the right economic perks in order to motivate people to work well. In a large company we must be very careful not to lose the individualized touch that is required in order to make employees feel appreciated.*

An ideal organizational work environment should facilitate opportunities for independence, recognition and responsibility among employees (Appelbaum, St-Pierre, & Glavas, 1998).

Manager A: *Work autonomy is associated with the level of maturity and work experience of everyone. …work autonomy becomes a key factor to become efficient and empowered by his/her boss. Although not all people are granted with work autonomy because either, they are not ready, or they are not interested, if an individual is a leader, he/she will seek for work autonomy if not granted...*

Leadership improves the organization's productivity only if there is an internal leadership control process. In other words, employees must be monitored with various measures for work effectiveness both individually and on teams (Brown & May, 2012).

Manager B: *To help a person become a leader, we must help him/her develop good communication skills, teach him/her to fulfill his commitments and become highly adaptable to the changes, and teach him/her how to manage work pressure.*

Team work effectiveness, level of entrepreneurship spirit and an established work performance criterion are the three key indicators of an appropriate leadership strategy. The work performance criterion should measure 1) standardized work structures that reflect the knowledge and use of processes, 2) trust and empathy development among team members, 3) improvement in self-esteem and entrepreneurship spirit as strengths for creativity, decision-making, and risk taken, and 4) team empowerment to make changes and work improvements (Carr, 1995).

Manager B: *A highly productive worker is a person able to perform the job function assigned to him/her integrating the necessary resources to achieve a goal... Somebody that is required to meet customer requirements, company budgets, product and performance quality, and comply with project management requirements...*

Somebody who has good communication skills, work discipline, is flexible in regard to working conditions, is able to handle work pressure successfully, and is commitment to meet goals... Must have resilience to face challenges and highly adaptable to daily work conditions.

It is necessary to analyze the organizational work environment and leadership system to understand the level of performance in the organization. The organizational work environment should be able to capture affective events that in turn influence employees' emotions as well as information processing behaviors which need to be supported by their leaders (Haakonsson, Burton, Obel, & Lauridsen, 2008).

Manager A: *If the company is putting work pressure on the employee, he/she must be able to tolerate it. If a person has self-control of his/her emotions, he/she can also control his relationship with others...*

Therefore, the organizational work environment and the leadership system must be perfectly aligned in order to obtain an adequate performance (Haakonsson et al., 2008). Leadership and administrative performance have a central role in the development of competitive advantages (Hall & Rowland, 2016).

Manager B: *A good leader has the power to persuade subordinates through communication, without this ability he cannot be an agent of change that is required during team activities.*

Current management thinking emphasizes the need of a clear strategy that recognizes the importance between leadership and performance. Furthermore, it is important to understand the difference between leadership and management in order to assume both roles effectively and facilitate the necessary conditions for employees to reach their best effort and remain motivated throughout the process (Hall & Rowland, 2016).

Manager A: *If a person knows what is expected of him/her as well as his/her limitations, he/she can decide to face the challenge...*

Manager B: *The group synergy is a consequence of good leadership, it helps subordinates to work with synergy and behave as a self-directed work team which helps supervisors by reducing their work load and face less problems...*

Upper management/company support is key for a worker to perform well. He/she should have the right support and guidance to achieve his/her goals and reach the desired levels of productivity.

5. GENERAL DISCUSSION

Companies seeking to increment their productivity levels are concerned about the influences on how salary inequalities impact in the individual productivity among middle management supervisors. For this reason, learning more about the relation among leadership, individual productivity, and wage compensation is paramount.

5.1. Leadership-Productivity Relation

This study had empirically tested the effects of leadership on the productivity levels among middle management employees using several human factors such as knowledge, personal skills, motivation, and work elements that shape employees character and allow them to positively impact their productivity levels. Recognition, autonomy, group cohesion, communication, pressure, and support represent the set of work elements that allow employees to expand their personal ability to contend with work changes and challenges (Lopez & Sahagun, 2016; Lopez & Sahagun, 2017; Lopez, Sahagun & Cisneros, 2017; Lopez, Sahagun & Ortiz, 2017; Lopez, Sahagun & Ortiz, 2018). Furthermore, organizations could increment work commitment among their employees through the implementation of leadership training which will have a positive and significant influence on the organizations' productivity levels. Therefore, making changes in work methods and establishing appropriate relationships and interactions between employees and their superiors will allow companies to improve their overall performance.

5.2. Moderating Effect of Wage Compensation in the Leadership-Productivity Relation

Research findings corroborate the moderating role of wage compensation in the leadership-productivity relation as proposed. Thus, inequality on wage compensation has a considerable effect on productivity outcomes and leadership behavior (Buttner & Lowe 2017).

Furthermore, perception of wage compensation inequality can be manifested in a form of justice and valuation of the responsibility for the fulfillment of company's goals and/or it can produce low levels of commitment between the worker and the organization. Thus, minimizing wage/salary inequality could boost productivity levels among employees. However, in order to make available additional monetary resources to close this wage/salary gap, organizations will have to learn how to work more efficiently.

5.3. Having an Internal Administrative Process to Control Leadership Levels Among Middle Management Supervisors

Majority of salary and wage inequality influence is via the interaction effects with leadership as a second order factor. This means that although middle management supervisors care about salary inequality, they mainly rely on their leadership role for the success of their team. However, preparing supervisors for their new role may not be enough to achieve the results desired by the organization. The organization must, under a transparent management process, control and develop leadership training programs and measure their effects among groups and supervisors until the desired leadership competence is achieved. This approach will facilitate the leadership management process.

Furthermore, salary alone does not guarantee neither high motivation levels among employees nor high achievement levels toward goals. Then, it is necessary to allow workers to integrate the internal factors and processes required to achieve their productivity goals. Simultaneously, companies must, through a combination of leadership training and salary compensation, incentivize work performance to achieve the desired productivity levels among employees.

6. THEORETICAL AND MANAGERIAL IMPLICATIONS

Although the notion of leadership-productivity is not new (Appelbaum, St-Pierre, & Glavas, 1998; Brown & Douglas, 2012; Carr, 1995; Haakonsson et al., 2008; Hall & Rowland, 2016; Kaplan & Norton, 2004), its treatment requires further empirical research in order to better understand and explain how salary inequality affects this relation (Schuck & Rabe-Hemp, 2018). It is important to discover the effects, direct and indirect, of perceived salary inequality in the leadership-productivity relation among middle management supervisors.

This research attempts to make a theoretical contribution in confronting the above issues in three ways. First, beyond corroborating various leadership factors suggested in previous studies (Lopez & Sahagun, 2016; Lopez & Sahagun, 2017; Lopez,

Sahagun & Cisneros, 2017; Lopez, Sahagun & Ortiz, 2017; Lopez, Sahagun & Ortiz, 2018), it provides an enriched and customized framework to fully understand the leadership effect on individual productivity among middle management production supervisors. Second, this framework enables researchers to identify the effects of perceived salary and wage inequality in the leadership-productivity relation among middle management supervisors the Mexican border. Third, determine the importance of developing an administrative process to control the negative impact that salary and wage inequality may have.

Thus, the findings of this research might well be important to upper managers interested in the establishment of productivity metrics associated with the leadership defined by the company, the supervisors' styles developed in internal programs, and the organizational learning process. Furthermore, companies can also establish development procedures for their supervision units through training focused on coaching and transformational leadership.

In sum, organizations are constantly changing, and it is necessary to be fully aware of the necessary resources to succeed in an ever-changing market. Thus, the benefits of this research can help managers determine which changes in their working methods are required to increment productivity. Furthermore, it could help designing training modules that help improve key work competencies among middle management employees and allow managers to act in a predictive manner in terms of the interests of their supervisors and positively influence their work satisfaction and development. As well as controlling the negative impact of perceived salary and wage inequality among them.

7. LIMITATIONS AND FUTURE RESEARCH

Some limitations should be taken into consideration when interpreting and drawing inferences based on the findings presented here. Neither the selection of participants nor the selection of the companies from which the participants were immersed was randomly performed. Thus, the sample might not be representative of the population. A second limitation is that the participants included were only people willing to participate. Such an approach limits the feasibility of estimating the non-response bias and testing for the differences between people who participated and people who did not participate in the study.

Furthermore, this research considered only production supervisors from one of the border countries (Mexico). Therefore, further research is needed using different types of supervisors and adding supervisors from the U.S border. Additionally, further research is needed to improve the understanding of this phenomenon under similar productivity and compensation levels.

REFERENCES

Al-Matari, M. & Al SwidiK & Fadsil B, (2014). The measurements of Firm Perfomance's Dimensions. *Asian Journal of Finance & Accounting*, *6*(1).

Appelbaum, S. H., St-Pierre, N., & Glavas, W. (1998). Strategic organizational change: The role of leadership, learning, motivation and productivity. *Management Decision*, *36*(5), 289–301. doi:10.1108/00251749810220496

Bozec, R., & Dia, M. (2007). Board structure and firm technical efficiency: Evidence from Canadian state-owned enterprises. *European Journal of Operational Research*, *177*(3), 1734–1750. doi:10.1016/j.ejor.2005.10.001

Brown, W., & May, D. (2012). Organizational change and development: The efficacy of transformational leadership training. *Journal of Management Development*, *31*(6), 520–536. doi:10.1108/02621711211230830

Busic, T., Robinson, L., & Ramburuth, P. (2010). Effects of leadership style on team learning. *Journal of Workplace Learning*, *22*(4), 228–248. doi:10.1108/13665621011040680

Buttner, E. H., & Lowe, K. B. (2017). Addressing internal stakeholders' concerns: The interactive effect of perceived pay equity and diversity climate on turnover intentions. *Journal of Business Ethics*, *143*(3), 621–633. doi:10.100710551-015-2795-x

Carr, J. T. (1995). Can we standardize leadership performance. *Management Development Review*, *8*(1), 25–27. doi:10.1108/09622519510771825

Chen, L., Gillenson, M. L., & Sherrell, D. L. (2002). Enticing online consumers: An extended technology acceptance perspective. *Information & Management*, *39*(8), 705–719. doi:10.1016/S0378-7206(01)00127-6

Destefanis, S., & Sena, V. (2007). Patterns of corporate governance and technical efficiency in Italian manufacturing. *MDE. Managerial and Decision Economics*, *28*(1), 27–40. doi:10.1002/mde.1310

Dzinic, J. (2015). *Correlation between the administrative leadership style and inclination towards organizational learning in local administrative organizations*. Ekon.

Fornell, C., & Larcker, D. F. (1981). Structural equation models with unobservable variables and measurement error. *JMR, Journal of Marketing Research*, *18*(February), 39–50. doi:10.1177/002224378101800104

Haakonsson, D. D., Burton, R. M., Obel, B., & Lauridsen, J. (2008). How failure to align organizational climate and leadership style affects performance. *Management Decision, 46*(3), 406–432. doi:10.1108/00251740810863861

Hair, J. F., Black, W. C., Babin, B. J., & Anderson, R. E. (2010). *Multivariate data analysis, a global perspective* (7th ed.). Prentice-Hall.

Hall, R. D., & Rowland, C. A. (2016). Leadership development for managers in turbulent times. *Journal of Management Development, 35*(8), 942–955. doi:10.1108/JMD-09-2015-0121

Kaplan, R. S., & Norton, D. P. (2004). The strategy map: Guide to aligning intangible assets. *Strategy and Leadership, 32*(5), 10–17. doi:10.1108/10878570410699825

Krol, M., Brouwer, W., & Rutten, F. (2013). Productivity costs in economic evaluations: Past, present, future. *PharmacoEconomics, 31*(7), 537–549. doi:10.100740273-013-0056-3 PMID:23620213

López-Vázquez,. F. & Sahagún, M. A. (2016). Estudio de Caso: Medicion de Habilidades Humanas y Clima Organizacional para Identificar y Desarrollar Acciones de Productividad Incremental. *20th Annual Western Hemispheric Trade Conference, Laredo, TX, USA*. Free Trade. http://freetrade.tamiu.edu/PDF/20Conf-Sessions.pdf#page=1

López-Vázquez, F., & Sahagún, M. A. (2017). *Análisis y Comprensión de la Relación Liderazgo Productividad en la Industria de Manufactura México-USA: Teoría y Prueba Empírica. II Congreso de Estudios de la frontera*, Brownsville, TX, USA.

López-Vázquez, F., Sahagún, M. A., & Cisneros-Lascurain, E. E. (2017). *Análisis de los principales factores de influencia en el desarrollo de productividad incremental: sector de manufactura electrónica y automotriz. 21th Annual Western Hemispheric Trade Conference*, Laredo, TX, USA.

López-Vázquez, F., Sahagún, M. A., & Ortiz-Rodriguez, F. (2017). Análisis y comprensión de factores que influyen en la productividad empresarial. *Congreso Internacional Multidiciplinario Veracruz 2017 Tecnologico Nacional de Mexico*. Instituto Tecnologico de Veracruz. Revista Impulso Tecnológico.

López-Vázquez, F., Sahagún, M. A., & Ortiz-Rodriguez, F. (2018). *Efectos entre Inequidad Salarial y Productividad de la Industria de Manufactura en la frontera México-Estados Unidos. 22th Annual Western Hemispheric Trade Conference*, Laredo, TX, USA.

McClelland, D. C., & Boyatzis, R. E. (1982). Leadership motive pattern and long-term success in management. *The Journal of Applied Psychology*, *67*(6), 737–743. doi:10.1037/0021-9010.67.6.737

Nicol, H. N., & Minghi, J. (2005). The continuing relevance of borders in contemporary contexts. *Geopolitics*, *10*(4), 680–687. doi:10.1080/14650040500436647

Olsen, J. A., & Richardson, J. (1999). Production gains from health care: What should be included in cost-effectiveness analyses? *Social Science & Medicine*, *49*(1), 17–26. doi:10.1016/S0277-9536(99)00116-1 PMID:10414837

Pan, Y., & Jackson, R. T. (2008). Ethnic difference in the relationship between acute inflammation and serum ferritin in US adult males. *Epidemiology and Infection*, *136*(3), 421–431. doi:10.1017/S095026880700831X PMID:17376255

Raja, A. S., & Palanichamy, P. (2011). Leadership Styles and Its Impact on Organizational Commitment. *The Journal of Commerce, 3*(4). Hailey College of Commerce, University of the Punjab, Pakistan.

Rappe, C., & Zwick, T. (2006). Developing leadership competence of production unit managers. *Journal of Management Development*, *26*(4), 312–330. doi:10.1108/02621710710740084

Rogerson, P. A. (2001). *Statistical methods for geography*. Sage. doi:10.4135/9781849209953

Sanchez, R. (2013). Reflecting on the Border. A Search for Alternative Research Approaches. EL colegio de la Frontera. *Frontera Norte*, *26*(3).

Sánchez Juárez, I. L., & Campos Benítez, E. (2010). Industria manufacturera y crecimiento económico en la frontera norte de México. *Región y Sociedad*, *22*(49), 45–89.

Schuck, A. M., & Rabe-Hemp, C. E. (2018). Investing in people: Salary and turnover in policing. *Policing*, *41*(1), 113–128. doi:10.1108/PIJPSM-09-2016-0137

Sheu, H. J., & Yang, C. Y. (2005). Insider Ownership Structure and Firm Performance: A productivity perspective study in Taiwan's electronics industry. *Corporate Governance*, *13*(2), 326–337. doi:10.1111/j.1467-8683.2005.00426.x

Sirmon, D. G., Hitt, M. A., & Ireland, R. D. (2007). Managing firm resources in dynamic environments to create value: Looking inside the black box. *Academy of Management Review*, *32*(1), 273–292. doi:10.5465/amr.2007.23466005

Werner, O., & Campbell, D. T. (1970). Translating, working through interpreters, and the problem of decentering. In R. Naroll & R. Cohen (Eds.), *A handbook of method in cultural anthropology* (pp. 398–420). Natural History Press.

Ziapour, A., Khatony, A., Kianipour, N., & Jafary, F. (2015). Identification and analysis of labor productivity components based on achieve model (case study: Staff of Kermanshah university of medical sciences). *Global Journal of Health Science*, *7*(1), 315. PMID:25560364

Chapter 13
The Work of the Committees in the Business Chambers to Promote the Development of Emerging Technologies

Samantha Ivett Méndez-Sánchez
INDEX, Mexico

ABSTRACT

Emerging technologies, understood as new technologies or the continuous development of existing technologies, have led to significant advances ranging from the simplification of everyday tasks to the automation and optimization of industrial processes; and therefore, to increase the productivity and quality of life of people. Who does not like to have more time to devote to other activities or to rest, or to have some device that allows one to simplify life with a click. Or in the case of companies, reduce production times, costs, and process failures. While we may think that responsibility for promoting the progress of emerging technologies; as well as the definition of public policies, regulations, and regulation of this matter lies in the figure of the government, the reality is that the work and interference of other actors such as the private sector is crucial to its progress.

DOI: 10.4018/978-1-6684-8088-5.ch013

INTRODUCTION

Even if there is no consensus on the definition of "emerging technologies" we can understand them as new technologies or the continuous development of existing technologies, which reach their productivity plateau from 10 to 15 years (Ortiz-Rodriguez at al. 2023), with a decade of maturation until they can penetrate the market on a large scale[1]. These new technological solutions have led to great advances, ranging from simplifying everyday tasks to automating and optimizing industrial processes and thus increasing the productivity and quality of life of individuals and businesses. Not to mention, they support the development of new products and services.

When talking about emerging technologies, the interesting thing is to limit their development under the progressive approach (Kraaijenbrink, 2022), of industry 5.0[2], which, while still considering productivity and efficiency as ends, aims to form a smarter society, ensuring social well-being, without neglecting the sustainability of the planet. The 3 pillars that this proposal considers are committed to focusing technological development on the human being, by promoting their talent and creativity in activities that allow them to devote their capacity to innovate; seek resilience with flexible and adaptable technologies, as well as with organizations that have the capacity to address a potential crisis in a timely manner, and thus ensure stable and sustainable performance; and finally, be sustainable to impact with policies that care for planetary boundaries (Konrad Lorenz University Foundation).

STATE OF THE ART

The (MIT Technology Review, 2023) gives a very practical overview of what it identifies as the 10 emerging technologies 2023 and which can help us visualize the advances in this area in our daily lives:[3]

CRISPR Against High Cholesterol

It is a genetic editing tool called "base editing" or "CRISPR 2.0" by which scientists can exchange one DNA base for another. Primary editing (also called "CRISPR 3.0") allows scientists to insert DNA fragments into a genome. If it works in people, it could allow scientists to replace disease-causing genes. All treatments are in an experimental section.

IA That Creates Images

Companies like OpenAI, Stability AI, Midjourney, Google have launched language models that convert text into image. When writing a brief description, the program is in charge of generating an image of what was requested immediately and today it has already advanced to converting text into videos, thinking that someday you could even create a movie by just entering the script into a computer. This technology is incorporated into commercial programs such as Photoshop.

The Chip Design That Changes Everything

RISC-V is an open standard that is changing the way companies create computer chips, as anyone can use it to design a chip at no cost. These chips have already started to be used on headphones, hard drives and artificial intelligence processors, and companies are also working on RISC-V designs for use in data centers or spacecraft.

Military Drones for Mass Use

Unfortunately, the technology also has harmful uses, such as drones that are used in warfare as a tactical advantage and for targeting civilians. The Ukrainian war brought to market low-budget drone models from China, Iran and Turkey that, while different from high-end precision attack Americans, add to their widespread use in warfare by changing the form of combat.

Abortive Telemedicine Pills

In 2021, during the pandemic, the U.S. Food and Drug Administration (FDA) temporarily allowed health care professionals to mail patients two abortion pills (mifepristone and misoprostol). Although it was later ruled that abortion was not a constitutional right, access to abortion pills remotely had already set a precedent. In January 2023, the FDA assured that it would allow online and physical retail pharmacies to dispense abortive pills to prescription patients.

Organs à la Carte

Genetic engineering can produce organs and prevent waiting deaths for transplants. The biotechnology industry is turning to gene editing to modify pigs, eliminating sugar molecules and adding other genes to make these animals look more human. However, in the future, organ engineering could completely dispense with animals. Research is already underway on how to design complex tissues from scratch.

The Inevitable Electric Vehicle

Sales of electric vehicles have increased significantly in 2023 over the previous year and it is also a fact that more and more governments are enacting emission reduction policies and people are becoming more aware of the care of the environment. As far as the European Union is concerned, the intention is that by 2035 combustion models will cease to be sold.

James Webb Space Telescope

NASA's James Webb Space Telescope (JWST), a collaboration between the United States, Europe and Canada, is the largest telescope ever sent to space. It is designed to detect infrared radiation, something that allows the cosmic dust to pass through. Astronomers hope it will allow them to find out how the first galaxies in the universe formed after the Big Bang. It could also provide information on the formation of the atmospheres of the planets of other solar systems, as well as witness the birth of new worlds.

Ancient DNA Analysis

There are new methods that make damaged DNA readable and are driving a boom in ancient DNA analysis. Ancient samples can shed light on mysteries in today's health.

Recycling of Batteries

Thanks to the advances in recycling that allow the old batteries to be converted into new ones and the usefulness of valuable metals recovered from old notebooks, corroded electric drills and electric vehicles to power tomorrow's cars, recycling takes on a leading role in meeting future needs and climate goals. The new methods allow recyclers to dissolve metals more effectively and separate them from waste cells. Currently, recycling facilities can recover almost all cobalt and nickel, also often aluminum, copper and graphite; and more than 80% of lithium used batteries and excess scrap from battery production.

In addition, and under the premise that emerging technologies also refer to the continuous development of existing technologies, cybersecurity stands out in this evolution, since for more than thirty years there has been talk of computer crime and more than twenty years ago the term cybercrime was coined. Defined as "the set of policy tools, concepts, actions, best practices and technologies that can be used to protect users, their devices and transmitted and/or stored information from security risks in the cyber environment"[4], cybersecurity has attracted the spotlight given the big effects on companies and individuals. According to a study by the UN Bureau of Crime and Drugs (UNODC), cybercrime at the international level has increased and surpassed conventional forms of crime. Simple and car theft harms about 2 percent of the population, while e-mail hacking harms 14 percent of the adult world population. And with data from Cybercrime Magazine, it is estimated that cybercrime losses were $6 trillion in 2021, and are expected to reach $10 trillion by 2025, representing a larger criminal market than the entire illegal drug business in the world.[5]

This scenario of emerging technologies tells us about an evolution to a reality that will bring great benefits, as well as challenges to humanity. That is why digital transformation, that is, the implementation of new technologies, is a core issue to bring to the industry agenda together with key players to be even more competitive internationally and to rise to the wave of the technological revolution under a sustainable approach. Hence, the relevance of talking about the interlocking channels offered by business chambers in Mexico for strategic issues.

THE ROLE OF BUSINESS CHAMBERS IN PUBLIC POLICY

Business chambers are institutions of public interest (Cardozo Brum, Myriam 2013) autonomous, with legal personality and their own heritage that represent, promote, and defend, nationally and internationally, the activities of industry, commerce, services, and tourism and collaborate with the government to achieve socio-economic growth, as well as the generation and distribution of wealth. They are also organizations of consultation and collaboration of the three levels of government, for the design, dissemination, and implementation of policies, programmed and instruments for the promotion of national economic activity. That is, the government should consult them on all matters related to the activities they represent.[6]

The business chambers, then, serve as consultation organizations for the government at the time of making public policies by being the representation of thousands of companies affiliated to them. This role is relevant because they have the pulse on different issues, which allows to have a clearer idea of the alternatives that are available to solve problems that, due to their relevance, are important to sector of society.

The interesting thing is that thanks to the work of the Mexican businessman, the challenges that come to be presented in different subjects under certain political, economic, social, and technological conditions have been placed on the table of the government to seek solutions or alternatives to safeguard the competitiveness of the industry and, therefore, for the country. Business chambers have influenced the issuance and amendment of decrees, laws, and regulations; proposed strategic collaboration schemes through framework collaboration agreements; as well as open channels of direct communication with companies to allow that information related to compliance with the government and share with the government the best practices developed by the industry.

THE WORK OF THE NATIONAL COMMITTEES

Within the structure of the national business chambers, the Committees acquire relevance by being the technical part in which the discussions, approaches, proposals and concerns that affect or promote the operations of the industry take place. The Committees are the ones that execute and implement the actions in different areas, serving as interlocutors and voice of the sector vis-à-vis the government.

Their work goes in two ways, whether the government issues any concerns that need to be discussed and analyzed in the committee to be transmitted to the companies or whether the companies are the ones who upload to the Committees the needs that are being presented at the local level.

The Committees constitute the main basis of operation in the Chambers, since, through them, their affiliated companies are integrated by areas of interest to jointly address problems, propose solutions and develop best practices for the benefit of the sector.

THE WORK OF THE NATIONAL INFORMATION TECHNOLOGY INDEX COMMITTEE

From my experience, being part of the coordination of the network of national committees in the business chamber representing the export manufacturing industry, **index**, I have been able to note the great role that the Information Technology Committee (IT) of this organization has had in the agenda of emerging technologies in Mexico, especially that related to cybersecurity.

The **Index** IT Committee made up of executives from companies with a long history, experts in the field who are part of prestigious firms, academics, and advisors, acquired an innovative approach to contribute to the development of the industry in information technologies.

In 2022, given the situation of cyber-attacks on both governments and large corporations, in October of that same year, as part of the analysis and research work, the Committee was given the task of launching a survey to **index** membership to know the situation of hacks in the manufacturing industry and be able to deal with this threat (Miro, 2012), where it identified that:

- 18% of the companies surveyed suffered at least one hack during 2022.
- The economic impact of hacks on industry reported in that period amounted to $2,300,000 dollars.
- Attacks were detected that were not carried out directly on the companies, but on their service providers, causing collateral damage to their operation.

One of the great conclusions on this issue was on the prevailing need for cyber security protocols, installed network security systems, certified corporate equipment and continuing education on these issues (Ortiz-Rodriguez, 2020).

The IT Committee therefore decided to undertake several actions during the period 2022-2023, namely:

- New survey to the index membership on cybersecurity in the industry in July 2023.
- Launch of an awareness and evangelization campaign to reduce the risks derived from hacking and cyber theft with informative capsules and circulars.
- Annual organization of an international cybersecurity forum inviting international organizations, federal government officials, national legislators, other business chambers, cybersecurity experts and academics to discuss and seek solutions in cybersecurity. Two editions have taken place so far.
- Index's active participation in the main national and international forums such as LATAM CISO 2023: Perspectives from industry leaders, for the discussion of opportunities to strengthen cybernetic cooperation between the United States and Mexico (March 2023); "Cybersecurity a shared responsibility" of the MX Internet Association to explore ways to contribute to a secure digital environment in Mexico (June 2023); "Cybersecurity in Mexico: situation, risks and challenges" of the Mexican Council of International Affairs where the urgency of having legislation that regulates cybersecurity in Mexico was made clear (June 2023), "Cooperation mechanisms for the support of cybersecurity systems" at the XIX Annual Meeting of the Business Security Front of the Colombian National Police, organized by the Mexican Association of Private Security Companies and BlackIND to exchange knowledge and good practices in the field business security (July 2023), among others.
- Synergy with key organizations for the development of Cybersecurity such as the Federal Telecommunications Institute, ANADIC MX, CANIETI, and academia such as Tecnológico de Monterrey and the Autonomous University of Tamaulipas.
- Incorporation of strategic affiliates, those specialized firms that strengthen the Committee's analysis and provide their expertise to the manufacturing industry.

- Creation of the index Industrial Cybersecurity Center for 2025. A non-profit organization from which innovation is made in terms of technological and cyber security protection and evolves to be able to offer index to the community, in principle and to society in general, an adequate level of stability so that their companies are not affected by cyber-attacks.
- Approach with the Federal Government to work on public policies on cybersecurity.

Delving into this last point, index, through the IT Committee, issued in November 2022 a position to the legislature on the need to create a Cybersecurity Law in Mexico.

First, arguing that the manufacturing industry considered the cybersecurity legislation and the issuance of the Cybersecurity Act to be extremely necessary, given the violations that companies in the industry have suffered and the effects they have brought to their operations, comparable to issues of heritage security. It was stated that the current challenges to keep information safe and the safe flow of business and economic transactions require, not only the private sector, but society in general, to seek the legal mechanisms that guarantee business continuity. So today more than ever, it was necessary to launch an awareness campaign to reduce the risks posed by cybercrime.

The index made the Committee of IT experts available to the legislature for any joint discussion or analysis on cybersecurity; suggested the creation of an Industrial Cybersecurity Innovation Center, with the participation of the public, private and academic sectors for its operation and development; supported the specialization and availability in the study plans of cybersecurity, as well as considering the allocation of items to address cybersecurity within government institutions.

This type of exercises by the private sector put on the table the discussion of the importance of having a regulatory framework for cybersecurity in Mexico, therefore, in April 2023, among other initiatives, the Federal Cybersecurity Law, on which index was also pronounced in July of that same year, after an internal discussion with a working group formed within the IT Committee for this purpose.

In this positioning, index welcomed the promulgation of a Federal Cybersecurity Law that seeks to regulate this matter, establish responsible institutions to observe and guarantee the cybersecurity of the country, and contemplate the persecution of cybercrime, establishing the rights and obligations of users in cyberspace under a Human Rights umbrella.

CONCLUSION

In this regard, and after making an analysis from the IT Committee, it was concluded that it is urgent to start with the implementation of legislations to advance the regulation of cybersecurity in Mexico. While it is difficult for a Law to cover "everything," it is necessary to start and then adjust on the fly.

Emphasis was placed on the fact that it is necessary to consider that there are jurisdictions behind the cyberspace infrastructure, so it is also necessary to keep in mind this matter as an aspect that transcends borders and that must be regulated by International Law. Although adherence to the Budapest Treaty remained a pending issue and perhaps it is now obsolete, it will be necessary to put on the table the fact of resuming an updated instrument from within the United Nations Organization and seeking the harmonization of laws internal to be able to adhere to and comply with it. Likewise, coexistence and harmony with other laws and international treaties on commercial matters, information protection, cybercrime and others will have to be contemplated.

These pronouncements by index, and by other business chambers, which are supported by research and analysis by their technical committees, are undoubtedly a great input for decision makers when making public policy on these issues. Although civil society, non-governmental organizations and academia are key actors, the role of business representation is added to these.

The committees, which are groups of people, who dedicate their time to collaborate and put their expertise at the service of the country, in the search for solutions when making public policy, an extremely admirable task if one considers that it is a pro bono activity.

While the technical part is fundamental to the understanding of emerging technologies, the political work of interlocking and representing the industry carried out by the business chambers on these issues is strategic to promote their development under a regulatory framework.

ACKNOWLEDGMENT

I want to give special thanks to Agustín Tiburcio, who has been a great mentor and guide for me during my professional development. Thank you for inviting me to collaborate as part of your team, it has been a great privilege.

REFERENCES

Brum, C. Myriam. (2013). Public policies: the debates of their analysis and evaluation. *Andamios, 10*(21), 39-59. https://www.scielo.org.mx/scielo.php?script=sci_arttext&pid=S1870-00632013000100003&lng=es&tlng=es

Kraaijenbrink, J. (2022). What is Industry 5.0 and how it will change companies. *Forbes*. https://forbes.es/empresas/167359/que-es-la-industria-5-0-y-como-cambiara-las-empresas/

Linares, M. (2012). *Cybercrime Phenomenology and criminology of crime in cyberspace*. Marcial Pons. https://soyancrig.com.gt/data/files/libros/el-cibercrimen.pdf

MIT Technology Review. (2023). The 10 Emerging Technologies. *MIT Technology Review*. https://www.technologyreview.es//listas/tecnologias-emergentes/2023

Office of the European Union. (2022). *Industry 5.0: Towards a sustainable, human-centric and resilient European industry*. Europea. https://op.europa.eu/en/publication-detail/-/publication/468a892a-5097-11eb-b59f-01aa75ed71a1/

Ortiz-Rodriguez, F., Barrera, R., Villanueva, V., Medina, M., & Gauna, A. (2020). APPVISO: citizen digital communication to prevent risk events in real time. In *The 21st Annual International Conference on Digital Government Research (dg.o '20)*. Association for Computing Machinery. 10.1145/3396956.3397019

Ortiz-Rodriguez, F., Tiwari, S., Sicilia, M. A., & Nikiforova, A. (2023). *Electronic Governance with Emerging Technologies*. First International Conference, EGETC 2022, Tampico, Mexico. https://www.konradlorenz.edu.co/blog/tecnologias-emergentes-y-su-impacto-en-las-organizaciones/

ENDNOTES

1 Digital SkillUp, What are emerging technologies and why are they relevant?, available at: https://www.digitalskillup.eu/es/articles/what-are-emerging-technologies/

2 Term coined in early 2021 by the European Commission. See Publications Office of the European Union, Industry 5.0: Towards a sustainable, human-centric and resilient European industry, available at: https://op.europa.eu/en/publication-detail/-/publication/468a892a-5097-11eb-b59f-01aa75ed71a1/

3 MIT Technology Review, The 10 Emerging Technologies 2023, available at: https://www.technologyreview.es//listas/tecnologias-emergentes/2023

4 Federal Telecommunications Institute, Glossary, available at: https://ciberseguridad.ift.org.mx/seccion/recomendaciones_generales

5 UNAM Gazette, Cyber Crimes on the Rise and with Exorbitant Numbers, available at: https://www.gaceta.unam.mx/delitos-ciberneticos-en-aumento-y-con-numeros-exorbitantes/#:~:text=Para%20la%20revista%20especializada%20Cybercrime,todo%20el%20negocio%20de%20drogas

6 Business Chambers and Confederations Act, available at: https://www.gob.mx/cms/uploads/attachment/file/663476/LCEC.pdf

Chapter 14
Transforming Learning and Remote Work:
Benefits of Technology Tools for Loyalty and Productivity

Susana Orallo Catoira
IRIUM, Mexico

ABSTRACT

Technology tools have played a key role in the transformation of learning and remote work, as well as in employee retention. Through access to information, improved communication, and optimized productivity, these tools have enabled organizations to meet the challenges of the digital age and flourish in an ever-changing environment. However, it is important to remember that technology is only a tool and that its true value lies in its effective integration into an overall strategy that promotes people's engagement and well-being. By embracing the potential of technology in a humanized way, organizations can build strong teams and successful projects in today's digital world.

1. INTRODUCTION

Today, we live in a world where technology has advanced significantly and has profoundly impacted our lives. From the way we learn, to how we work and interact, technology has become increasingly present and relevant. In this chapter, we will explore in detail how technological tools have transformed learning and remote work, and how they have helped to avoid staff turnover and retain people on projects.

DOI: 10.4018/978-1-6684-8088-5.ch014

As we move into a digitally connected society, it is essential to understand how these tools have facilitated learning, optimized remote work, and enabled organizations to thrive in this constantly evolving environment. However, it is also significant to remember that technology is only a tool, and that success lies in its effective integration into an overall strategy that promotes people's engagement and well-being.

Very importantly, throughout this chapter we will talk about technological tools, but without mentioning any of them. Nor will I talk about the value that the use of Big Data, Artificial Intelligence, Machine Learning, the Internet of Things, and other types of technologies that are strategic for the evolution of companies can bring to organizations. The aim of this text is not to delve into technologies but to create a symbiosis between technology and people.

Regarding the different solutions or tools that I will be mentioning in the text, there are different products or solutions known in the environment that are suitable for each of the points that are presented below. The purpose of this text is not to highlight or mention any manufacturer, but to emphasize the application of technological solutions with a single purpose, to improve the quality of life of our employees.

In case any reader is interested in knowing my professional opinion about any of the proposals presented, I will be pleased to analyze your needs and recommend, based on my experience, the most suitable manufacturer's solution for your requirements.

2. FACILITATING LEARNING

In the Area of Learning, a true revolution and evolution have been experienced thanks to technological tools, which have facilitated access to information and online interaction. In the following, we will explore how these tools have transformed learning in various areas:

2.1. Online Education and e-Learning

Online education and e-learning have opened the doors to global access to education, overcoming geographic and economic barriers. No matter where you are physically located, you can access prestigious educational programs from anywhere in the world. These tools also offer personalization and flexibility, adapting learning to the individual needs of students. Interactivity and collaboration are other fundamental aspects, as they encourage active participation and the collective construction of knowledge. In addition, assessment and feedback tools allow for more effective progress monitoring and continuous learning.

2.2. Corporate Learning and Online Training

In the corporate environment, technological tools have revolutionized the way training and professional development are delivered. These tools allow for a significant reduction in costs and time by eliminating the need to travel or reserve space for training. They also ensure that employees are up to date in their field, providing them with access to up-to-date information quickly and conveniently. Gamification and motivation also play an important role, as they make learning more attractive and entertaining. In addition, technological tools allow for more effective performance tracking and measurement, making it easier to assess the impact of training on business results.

3. OPTIMIZING REMOTE WORK

Although the possibility of companies having a "Remote Work" modality has been legally studied and implemented for more than 40 years, it was only after the Covid-19 pandemic that it became a reality. The fact of being forced to work from home, because of not being able to go to our work centers, led, in record time, to adapt operations, systems, policies, procedures, ways of managing, leading, and even reforming laws to meet and correctly cover this new reality.

Here we will highlight the benefits that these tools have brought to remote work:

3.1. Online Communication and Collaboration

Technological tools have facilitated online communication and collaboration, bringing geographically dispersed teams closer together. Video calls and virtual conferences allow team members to interact as if they were in the same physical location, improving communication and strengthening human bonds. In addition, instant messaging platforms foster real-time communication and project collaboration, streamlining workflow and promoting the active participation of all team members. Virtual workspaces are also valuable tools, facilitating document co-editing and efficient project management.

3.2 Remote Access to Tools and Data

Technology has enabled remote access to tools and data through various solutions. Cloud storage is one of the most prominent, as it allows access to files and documents from anywhere with an Internet connection. This capability is especially relevant for remote work, as employees can access the necessary resources without geographical restrictions. In addition, desktop virtualization has provided employees with a complete work environment, regardless of their physical location, enabling them to perform their tasks efficiently and productively. Finally, data security is a crucial aspect of remote work, and technological tools have provided reliable solutions to protect sensitive information and ensure data confidentiality.

3.3. Productivity and Labor Flexibility

Remote work has provided people with greater flexibility in managing their time and has allowed them to reconcile work and personal responsibilities more effectively. By not having to spend time commuting to the office, employees gain more hours for family and personal reconciliation, allowing them to perform more activities and maintain a healthy balance in their lives. Time management tools are important allies in organizing tasks and staying focused on goals. In addition, monitoring and performance metrics provide a clear view of individual and team performance, allowing you to balance productivity with work-life balance and making it easier to identify areas for improvement.

I must make a special mention in this section, the fact that, from the companies, we must ensure to maintain a balance between work and personal life of employees who work from home. When we talk about Digital Disconnection, we talk about respecting the time and space of our employees, the fact that they are at home working does not imply that they are more available to the company. Through training we must help them to understand that they must manage their time correctly and that the commitment must be bidirectional; from the company, they cannot be contacted outside the established working hours; from home, the worker will not be able to dedicate working time to other activities. This learning must be on both sides, only in this way can a balance be achieved.

4. AVOIDING STAFF TURNOVER AND KEEPING PEOPLE LOYAL

Technology has also contributed to the loyalty of people to projects. It is not a matter of "retaining" talent, but of making people feel satisfied with the assignment and conditions that organizations offer them. It is not retention; it is voluntariness and enthusiasm to stay in the project they are offered.

In this aspect, not only technology is important, but the company has a great responsibility to develop and implement an organizational culture that stands out for values that are easily identifiable for workers. By making them feel that they are part of its creation and that different perspectives have been considered with a common purpose.

Technology provides organizations with the tools to transfer this Organizational Culture to all their collaborators and to emphasize all the actions inherent to it.

Here we will highlight which of these technological tools have influenced job satisfaction and commitment to the organization:

4.1 Communication and Participation

Corporate social networks and intranets have created a virtual community within organizations, facilitating interaction among employees. Despite not working in the same physical space, these tools create a sense of belonging to a project and promote the company's values. In addition, recognition platforms promote open communication and recognition of a job well done, strengthening motivation and a sense of achievement. Transparency and access to relevant information empower employees by providing them with a clear vision of the project and the organization, which contributes to their sense of belonging and active participation.

4.2 Professional Development and Continuous Learning

Technological tools offer numerous opportunities for professional development and continuous learning. Internal training platforms provide employees with the possibility to grow and develop within the organization, which strengthens their sense of advancement and contributes to employee loyalty. Likewise, external training platforms allow employees to stay current in their fields and acquire relevant new skills, many with official certifications to support their learning. In addition, superior online training, such as bachelor's and master's degree options, provides valuable opportunities for employees' academic and professional growth.

4.3 Organizational Culture and Wellness

Technology can also be an ally in promoting a healthy organizational culture and wellness. Wellness and mental health tools can promote balance and well-being in the work environment, contributing to employee satisfaction and retention. These tools can range from incentives to encourage physical exercise, to the use of telemedicine tools to achieve a democratization of health and a decrease in absenteeism due to medical visits. Special mention should be made of psychological support and coaching.

In addition, financial health tools can help and accompany workers in their search for economic stability and a real adjustment of their income and expenses, which contributes to their overall well-being.

Virtual spaces for sports, social and recreational activities are another way of using technology to promote interaction and relationship building beyond work, which strengthens the bonds between team members and improves the work climate.

Finally, flexibility and work-life balance are aspects valued by employees, and technology can be adapted to their individual needs, allowing the customization of work schedules and policies to achieve a balance between productivity and work-life balance.

In this section, it is very important to emphasize that these tools, which are extra benefits for workers, are 100% inclusive. While the training and production and work tools are almost completely adapted to be used by people with disabilities, in the case of work benefits tools, there are not so many that meet this need. The fact that a company implements work-benefit solutions, without considering this aspect, automatically makes the company is exercising discrimination. That is why I emphasize this aspect.

5. CONCLUSION

Technology tools have played a key role in the transformation of learning and remote work, as well as in employee retention. Through access to information, improved communication, and optimized productivity, these tools have enabled organizations to meet the challenges of the digital age and flourish in an ever-changing environment. However, it is important to remember that technology is only a tool and that its true value lies in its effective integration into an overall strategy that promotes people's engagement and well-being.

By embracing the potential of technology in a humanized way, organizations can build strong teams and successful projects in today's digital world. But how can this be achieved? The answer lies in understanding that technology cannot replace the importance of a strong organizational culture and ingrained ethical values. Rather, technology can be a powerful ally in driving engagement and job satisfaction. In this regard, organizations must use technology responsibly and consciously. Making employees aware of their individual and team responsibilities is fundamental to achieving a harmonious and satisfying work environment.

The effective integration of technology implies always considering the human impact and promoting a balance between the use of digital tools and the emotional and social needs of individuals. This is the only way to achieve a true humanization of technology.

ACKNOWLEDGMENT

Dedicated to my mother, who taught me the value of work and effort, without ever losing the illusion of enjoying the process.

ADDITIONAL READING

Brown, W., & May, D. (2012). Organizational change and development: The efficacy of transformational leadership training. *Journal of Management Development, 31*(6), 520–536. doi:10.1108/02621711211230830

Busic, T., Robinson, L., & Ramburuth, P. (2010). Effects of leadership style on team learning. *Journal of Workplace Learning, 22*(4), 228–248. doi:10.1108/13665621011040680

Jabbar, M. A., Tiwari, S., & Ortiz-Rodriguez, F. (2023). Smart Urban Computing Applications. River Publishers

Medina-Quintero, J. M., Sahagun, M. A., Alfaro, J., & Ortiz-Rodriguez, F. (2023). Global Perspectives on the Strategic Role of Marketing Information Systems. *Advances in Marketing, Customer Relationship Management, and E-Services, 2023*, 1–387. doi:10.4018/978-1-6684-6591-2

Ortiz-Rodriguez, F., Tiwari, S., Sicilia, M. A., & Nikiforova, A. (2023). Electronic Governance with Emerging Technologies. *First International Conference, EGETC 2022*, Tampico, Mexico.

Quintero, M., Melchor, J., & Echeverria, R. Osiris & Ortiz-Rodriguez, F. (2021). *Confianza y calidad de la información para la satisfacción y lealtad del cliente en el e-Banking con el uso del teléfono celular.* Contaduría y Administración. doi:10.22201/fca.24488410e.2022.2965

Schuck, A. M., & Rabe-Hemp, C. E. (2018). Investing in people: Salary and turnover in policing. Policing. *Policing, 41*(1), 113–128. doi:10.1108/PIJPSM-09-2016-0137

Werner, O., & Campbell, D. T. (1970). Translating, working through interpreters, and the problem of decentering. In R. Naroll & R. Cohen (Eds.), *A handbook of method in cultural anthropology* (pp. 398–420). Natural History Press.

APPENDIX

Business Case

PROJECT "JUNIORS" || HIGH AVAILABILITY SERVICE BANKING SECTOR - MEXICO TWENTY20

As an example of the need to humanize the technological environment we face. I will present a case, which to my satisfaction was a great success, and which I developed in recent years.

At the end of 2019, and from the hand of one of the largest technology integrators that exist worldwide, we received a very specific requirement and that, as an end customer, had one of the most representative banks in Mexico.

Need of the final client:

To implement technological solutions for monitoring and control of activity. The objective was to have enough information to be able to perform an analysis on it, and to be able to apply automation that could optimize the processes and make the service more profitable.

Definition of the service to be applied:

High availability service, 24x7, attention to several countries, critical service for the entity since it is the one that manages all the banking transactions of an app, offices, web, and ATMs.

Problem:

Although the technology implementation part of the service was progressing favorably, two factors were negatively influencing the entire progress of the project.

- Generational and technological change: One of the technological areas of the service was being provided by professionals of retirement age, and within the team, there were no people prepared to take over from the next generation.
- Turnover: Due to the intensity of the work for the need of the service itself, a very high turnover rate was generated, which had a significant impact on the results of the service, both in terms of KPIs and in terms of costs due to loss of operation.

This led us to develop a strategy outside of technology, and within more human aspects, we would not succeed in the project, if we did not get the stabilization of the team, and we solved the situation of critical positions not having replacements.

From the company we developed a project called JUNIORS, it was a 12-month plan, in which the objective was to incorporate 35 junior professionals, to stabilize 30. With them, we would be able to generate a base layer in the service that would allow collaborators who were already stabilized in the service to move up a level in knowledge to cover those critical positions that were being executed by people of retirement age.

It was decided to hire 35 people, knowing that in the first two to three months it was very likely that 30 of those 35 would end up remaining due to voluntary departures, which was foreseen. The real need was to stabilize 30 employees.

The project included a career plan, not only to accompany them throughout those 12 months with technical training but also to reinforce them with training in soft skills, seeking to ensure that, in that first year of service, the employees would also mature on a personal level. The plan also included several recreational activities to foster cohesion, teamwork, and a sense of belonging.

The problem was that the start date was set for March 2020, which was precisely the month in which the pandemic broke out and we all went home.

If the service was already affected by turnover, with this new element, it suffered even more.

Thanks to technology and the tools it puts in our hands, the project was started 100% remotely, in May 2020. We had to redesign the entire plan that had been developed, it was a strong challenge to get the employees, who did not know us in person, who had not been in our offices, nor in those of the end customer, who had never worked in a remote mode, to get involved in the project in the way we needed.

And yes, finally it was a success, but it was a success because we applied a very strong human component in all the steps we took, if we eliminate human contact, and we only have a screen in the middle, we must strive much more to apply humanity to the activity.

The JUNIORS project ended in June 2021, we did it with a great Virtual Graduation Party, with 32 of the 35 collaborators who started a year before, and with a great congratulation from the final client.

The two proposed objectives were achieved, but above all, a very important improvement in service quality levels was achieved.

I will say that it was not until September of that same year that we were able to personally meet each of the colleagues that made up this group, and it was very satisfying to receive them.

Technology yes, but humanity x2

Compilation of References

Ahmad, A., Maynard, S. B., Desouza, K. C., Kotsias, J., Whitty, M. T., & Baskerville, R. L. (2021). How can organizations develop situation awareness for incident response: A case study of management practice. *Computers & Security, 101*, 102122. doi:10.1016/j.cose.2020.102122

Ahmadi, A., Moradi, M., Cherifi, C., Cheutet, V., & Ouzrout, Y. (2018). Wireless connectivity of CPS for smart manufacturing: A survey. *2018 12th International Conference on Software, Knowledge, Information Management & Applications (SKIMA)*, (pp. 1–8). ACM.

Ahmadi, A., Sodhro, A. H., Cherifi, C., Cheutet, V., & Ouzrout, Y. (2019). Evolution of 3C cyber-physical systems architecture for industry 4.0. *Service Orientation in Holonic and Multi-Agent Manufacturing: Proceedings of SOHOMA 2018*, (pp. 448–459). ACM.

Al-Ali, A. R., Gupta, R., & al Nabulsi, A. (2018). Cyber physical systems role in manufacturing technologies. *AIP Conference Proceedings, 1957*, 050007. doi:10.1063/1.5034337

Alhusain, S., Coupland, S., John, R., & Kavanagh, M. (2013). Towards machine learning based design pattern recognition. In *2013 13th UK Workshop on Computational Intelligence (UKCI)*. IEEE. 10.1109/UKCI.2013.6651312

Al-Matari, M. & Al SwidiK & Fadsil B, (2014). The measurements of Firm Perfomance's Dimensions. *Asian Journal of Finance & Accounting*, *6*(1).

Al-Qahtani, A. F., & Cresci, S. (2022). The COVID-19 scamdemic: A survey of pishing attacks and their countermeasures during COVID-19. *IET Information Security, 16*(5), 324–345. doi:10.1049/ise2.12073 PMID:35942004

Al-Shanfari, I., Yassin, W., Tabook, N., Ismail, R., & Ismail, A. (2022). Determinants of Information Security Awareness and Behaviour Strategies in Public Sector Organizations among Employees. *International Journal of Advanced Computer Science and Applications, 13*(8). doi:10.14569/IJACSA.2022.0130855

Anwar, M., He, W., Ash, I., Yuan, X., Li, L., & Xu, L. (2017). *Gender difference and employees' cybersecurity behaviors*. Comptuters in Human Behavior., doi:10.1016/j.chb.2016.12.040

Appelbaum, S. H., St-Pierre, N., & Glavas, W. (1998). Strategic organizational change: The role of leadership, learning, motivation and productivity. *Management Decision*, *36*(5), 289–301. doi:10.1108/00251749810220496

Aria, M., & Cuccurullo, C. (2017). Bibliometrix: An R-tool for comprehensive science mapping analysis. *Journal of Informetrics*, *11*(4), 959–975. doi:10.1016/j.joi.2017.08.007

Ariga, Y. M. (1999). Microstructural Change and Hardness of Lead Free Solder Alloys. Dept. of Metallurgical Engineering, School of Engineering, Tokai University.

Athavale, S., & Balaraman, V. (2013). Human behavioral modeling for enhanced software Project Management. In *7th International Conference on Software Engineering*, (pp. 15–7). IEEE.

Atwah, A. M. (2022). Investigating the drivers of cybersecurity enhancement in public organizations: The case of Jordan. *The Electronic Journal on Information Systems in Developing Countries*, *88*(5), 12223. doi:10.1002/isd2.12223

Baldwin, R. E. (2012). *Global supply chains: why they emerged, why they matter, and where they are going.*

Baluch, A. (2023, June 18). Best ERP systems of 2023. *Forbes.* https://www.forbes.com/advisor/business/software/best-erp- systems/

Berg, M. (2005). *The age of manufactures, 1700-1820: Industry, innovation and work in Britain.* Routledge. doi:10.4324/9780203990971

Berlilana, N., Noparumpa, T., Ruangkanjanases, A., Hariguna, T., & Sarmini. (2021). Organization benefit as an outcome of organizational security adoption: The role of cyber security readiness and technology readiness. *Sustainability (Basel)*, *13*(24), 13761. doi:10.3390u132413761

Bigliardi, B., Bottani, E., & Casella, G. (2020). Enabling technologies, application areas and impact of industry 4.0: A bibliographic analysis. *Procedia Manufacturing*, *42*, 322–326. doi:10.1016/j.promfg.2020.02.086

Bishop, E. (2023). What is manufacturing ERP (Enterprise Resource Planning)? *Market Business News.* https://marketbusinessnews.com/what-is-manufacturing-erp-enterprise-resource-planning/289612/

Blecker, R. A. (2014). The Mexican and US economies after twenty years of nafta. *International Journal of Political Economy.* https://www.jstor.org/stable/24696555

Blythe, J. M., & Coventry, L. (2018). Costly but effective: Comparing the factors that influence employee anti-malware behaviours. *Computers in Human Behavior*, *87*, 87–97. doi:10.1016/j.chb.2018.05.023

Boehm, B., & Sullivan, K. (2000). *Software Economics: A Roadmap.* . doi:10.1145/336512.336584

Boglioli, D. (2023). *EBOOK-MFG-Improving-Op-Res-8.5x11-001*. Nozomi Networks. https://www.nozominetworks.com/downloads/US/Nozomi-Networks-Manufacturing-Improving-Operational-Resilience-eBook.pdf

Boglioli, D. (2023). *The Challenges of the Vulnerability Assessment Process in ICS Space*. Nozomi Networks. https://www.nozominetworks.com/blog/the-challenges-of-ics-vulnerability-assessment-process/

Bokstein, B. S. A. O. (2011). Gain Boundary Difussion,Stresses and segregation. Switzerland: Trans Tech Publications.

Boyson, S. (2014). Cyber supply chain risk management: Revolutionizing the strategic control of critical IT systems. *Technovation*, *34*(7), 342–353. doi:10.1016/j.technovation.2014.02.001

Bozec, R., & Dia, M. (2007). Board structure and firm technical efficiency: Evidence from Canadian state-owned enterprises. *European Journal of Operational Research*, *177*(3), 1734–1750. doi:10.1016/j.ejor.2005.10.001

Brown, W., & May, D. (2012). Organizational change and development: The efficacy of transformational leadership training. *Journal of Management Development*, *31*(6), 520–536. doi:10.1108/02621711211230830

Brum, C. Myriam. (2013). Public policies: the debates of their analysis and evaluation. *Andamios, 10*(21), 39-59. https://www.scielo.org.mx/scielo.php?script=sci_arttext&pid=S1870-00632013000100003&lng=es&tlng=es

Busic, T., Robinson, L., & Ramburuth, P. (2010). Effects of leadership style on team learning. *Journal of Workplace Learning*, *22*(4), 228–248. doi:10.1108/13665621011040680

Buttner, E. H., & Lowe, K. B. (2017). Addressing internal stakeholders' concerns: The interactive effect of perceived pay equity and diversity climate on turnover intentions. *Journal of Business Ethics*, *143*(3), 621–633. doi:10.100710551-015-2795-x

Byle, F. J. (2006). A Study of Copper Dissolution in Pb-Free Solder Fountain Systems. *PanPacific Microelectronics Symposium.*

Caldarulo, M., Welch, E. W., & Feeney, M. K. (2022). Determinants of cyber-incidents among small and medium US cities. *Government Information Quarterly*, *39*(3), 101703. doi:10.1016/j.giq.2022.101703

Caldas, R. D., Rodrigues, A., Gil, E. B., Rodrigues, G. N., Vogel, T., & Pelliccione, P. (2020). A hybrid approach combining control theory and AI for engineering self-adaptive systems. In *Proceedings of the IEEE/ACM 15th International Symposium on Software Engineering for Adaptive and Self-Managing Systems*, (pp. 9-19). IEEE. 10.1145/3387939.3391595

Campos, R., & Monroy, L. (2016). La relación entre crecimiento económico y pobreza en México. *Investigacion Economica*, *75*(298), 77–113. doi:10.1016/j.inveco.2016.11.003

Carrapico, H., & Farrand, B. (2020). Discursive continuity and change in the time of Covid-19: The case of EU cybersecurity policy. *Journal of European Integration*, *42*(8), 1111–1126. doi: 10.1080/07036337.2020.1853122

Carr, J. T. (1995). Can we standardize leadership performance. *Management Development Review*, *8*(1), 25–27. doi:10.1108/09622519510771825

Catarci, T., Firmani, D., Leotta, F., Mandreoli, F., Mecella, M., & Sapio, F. (2019). A conceptual architecture and model for smart manufacturing relying on service-based digital twins. *Proceedings - 2019 IEEE International Conference on Web Services, ICWS 2019 - Part of the 2019 IEEE World Congress on Services*, (pp. 229–236). IEEE. 10.1109/ICWS.2019.00047

Chen, G., Wang, P., Feng, B., Li, Y., & Liu, D. (2020). The framework design of smart factory in discrete manufacturing industry based on cyber-physical system. *International Journal of Computer Integrated Manufacturing*, *33*(1), 79–101. doi:10.1080/0951192X.2019.1699254

Chen, H. (2017). Theoretical Foundations for Cyber-Physical Systems: A Literature Review. *Journal of Industrial Integration and Management*, *02*(03), 1750013. doi:10.1142/S2424862217500130

Chen, L., Gillenson, M. L., & Sherrell, D. L. (2002). Enticing online consumers: An extended technology acceptance perspective. *Information & Management*, *39*(8), 705–719. doi:10.1016/S0378-7206(01)00127-6

Collobert, R., Weston, J., Bottou, L., Karlen, M., Kavukcuoglu, K., & Kuksa, P. (2011). Natural language processing (almost) from scratch. *Journal of Machine Learning Research*, *12*(Aug), 2493–2537.

Corallo, A., Lazoi, M., Lezzi, M., & Pontrandolfo, P. (2021). Cybersecurity challenges for manufacturing systems 4.0: Assessment of the business impact level. *IEEE Transactions on Engineering Management*.

Craigen, D., Diakun-Thilbault, N., & Purse, R. (2014). Defining cybersecurity. *Technology Innovation Management Review*, *4*(10), 13–21. doi:10.22215/timreview/835

Dambra, S., Bilge, L., & Balzarotti, D. (2020). SoK: Cyber Insurance – Technical Challenges and a System Security Roadmap. *2020 IEEE Symposium on Security and Privacy (SP)*, San Francisco, CA, USA. 10.1109/SP40000.2020.00019

David Hillman, R. W. (2018). Dissolution Rate of Specific Elements in SAC305 Solder. *Proceedings of SMTA International*.

David, R., & Anderson, D. J. (2012). *Estadistica para negocios y economia*. Cengage Learning.

Dawson, J., & Thomson, R. (2018). The future cybersecurity workforce: Going beyond technical skills for successful cyber performance. *Frontiers in Psychology*, *9*, 744. doi:10.3389/fpsyg.2018.00744 PMID:29946276

Destefanis, S., & Sena, V. (2007). Patterns of corporate governance and technical efficiency in Italian manufacturing. *MDE. Managerial and Decision Economics*, *28*(1), 27–40. doi:10.1002/mde.1310

Di Maio, D. B. W. (2008). Measurements of Copper Dissolution in lead Free Solder Alloys. United kingdom: National Physical Laboratory.

Ding, K., Lei, J., Zhang, F., Wang, Y., & Wang, C. (2020). Analyzing the cyber-physical system–based autonomous collaborations among smart manufacturing resources in a smart shop floor. *Proceedings of the Institution of Mechanical Engineers. Part B, Journal of Engineering Manufacture*, *234*(3), 489–500. doi:10.1177/0954405419875340

Drápala, P. K. (2012). *Study of Interaction Between.* Salford University.

Duggal, A. S., Malik, P. K., Gehlot, A., Singh, R., Gaba, G. S., Masud, M., & Al-Amri, J. F. (2022). A sequential roadmap to Industry 6.0: Exploring future manufacturing trends. *IET Communications*, *16*(5), 521–531. doi:10.1049/cmu2.12284

Dzinic, J. (2015). *Correlation between the administrative leadership style and inclination towards organizational learning in local administrative organizations*. Ekon.

Dzyana, H., Pasichnyk, V., Garmash, Y., Naumko, M., & Didych, O. (2022). The System for Ensuring the Information Security of the Organization in the Context of Covid-19 Based on Public-Private Partnership. *International Journal of Computer Science and Network Security*, 19-24. doi:10.22937/IJCSNS.2022.22.6.4

Egghe, L. (2006). Theory and practise of the g-index. *Scientometrics*, *69*(1), 131–152. doi:10.100711192-006-0144-7

Escobar, A. (2011). Determinantes del empleo en la industria manufacturerar en México. *Papeles de Población*, 251–276.

EVS. (2000). *EVS International Recovery solder*. Obtenido de Leaders on Recovery Solder. http://www.solderrecovery.com/spanish/sq4.html

FBI. (2021). *Internet Crime Complaint Center IC3*. FBI. https://www.ic3.gov/

Fenton, N., Hearty, P., Neil, M., & Radlinski, L. (2010). Software project and quality modelling using Bayesian networks. In *Artificial intelligence applications for improved software engineering development: New prospects* (pp. 1–25). IGI Global. doi:10.4018/978-1-60566-758-4.ch001

Fernandez, R., Staff, T., Crouse, M., Partida, D., Corrales, E., Ayuya, C., & Kaelin, M. W. (2022, December 6). IOT tech expo global 2022: Exploring Satellite and telco connectivity for businesses. *TechRepublic*. https://www.techrepublic.com/article/iot-tech-expo-global-satellite-telco- connectivity/?utm_source=email&utm_medium=referral&utm_campaign=top-story-of-the-day

Fielder, A., Panaousis, E., Malacaria, P., Hankin, C., & Smeraldi, F. (2016). Decision support approaches for cyber security investment. *Decision Support Systems, 86*, 13-23. doi:10.1016/j.dss.2016.02.012

Filieri, A., Hoffmann, H., & Maggio, M. (2015). Automated multi-objective control for self-adaptive software design. *Proceedings of the 2015 10th Joint Meeting on Foundations of Software Engineering*. 10.1145/2786805.2786833

Firdaus, R., Xue, Y., Gang, L., & Ali, M. S. (2022). Artificial Intelligence and Human Psychology in Online Transaction Fraud. *Frontiers in Psychology*, *13*, 947234. Advance online publication. doi:10.3389/fpsyg.2022.947234 PMID:36304851

Fletcher, R. (2021). *Manufacturing Process Innovation for industrials Manufacturing*. McKinsey.

Fornell, C., & Larcker, D. F. (1981). Structural equation models with unobservable variables and measurement error. *JMR, Journal of Marketing Research*, *18*(February), 39–50. doi:10.1177/002224378101800104

Frontoni, E., Loncarski, J., Pierdicca, R., Bernardini, M., & Sasso, M. (2018). Cyber physical systems for industry 4.0: towards real time virtual reality in smart manufacturing. *Augmented Reality, Virtual Reality, and Computer Graphics: 5th International Conference, AVR 2018, Otranto, Italy, June 24–27, 2018. Proceedings*, *5*(Part II), 422–434.

Furnell, S., Haney, J., & Theofanos, M. (2021). Pandemic Parallels: What can cybersecurity learn from Covid-19? *Computer*, *54*(3), 68–72. doi:10.1109/MC.2020.3046888 PMID:34385715

Gao, H., Peng, Y., Jia, K., Wen, Z., & Li, H. (2015). Cyber-Physical Systems Testbed Based on Cloud Computing and Software Defined Network. In *2015 International Conference on Intelligent Information Hiding and Multimedia Signal Processing (IIH-MSP)*. IEEE. 10.1109/IIH-MSP.2015.50

García, G., & García, G. (2020). Polarización salarial y ocupacional en México. *Estudios Económicos*, 153-188.

George, A. S., & George, A. H. (2020). Industrial revolution 5.0: The transformation of the modern manufacturing process to enable man and machine to work hand in hand. *Journal of Seybold Report ISSN*, (1533), 9211.

Ghezzi, C., Jazayeri, M., & Mandrioli D. (2002). *Fundamentals of software engineering*. Prentice Hall PTR.

Gobierno del Estado de Tamaulipas. (2017). *Plan Estatal de Desarrollo 2016-2022*. Victoria, Tamaulipas, México.: Gobierno del Estado de Tamaulipas.

Gómez-Rodríguez, T., Ríos-Bolivar, H., & Aali-Bujari, A. (2018). Salario eficiente y crecimiento económico para el caso de América LAtina. *Ensayos*, 213-235.

Goro Izuta, T. T. (2009). Simplified Measuring Method of Copper Concentration in solder bath utilizing copper disolution. *IEEE Transactions on Electronics Packaging Manufacturing*, *32*(3), 138–143. doi:10.1109/TEPM.2009.2021259

Goyanes, M., & Demeter, M. (2020). How the geographic diversity of editorial boards affects what is published in JCR-Ranked communication journals. *Journalism & Mass Communication Quarterly*, *97*(4), 1–26. doi:10.1177/1077699020904169

Gross, J. (2022). *Leverage ERP to drive lean manufacturing initiatives*. Pemeco Consulting. https://www.pemeco.com/leverage-erp-to-drive-lean-manufacturing-initiatives/

Gujarati, N., & Porter, D. (2010). *Econometría*. McGraw-Hill.

Gunes, V., Peter, S., Givargis, T., & Vahid, F. (2014). A survey on concepts, applications, and challenges in cyber-physical systems. [TIIS]. *KSII Transactions on Internet and Information Systems*, *8*(12), 4242–4268.

Haakonsson, D. D., Burton, R. M., Obel, B., & Lauridsen, J. (2008). How failure to align organizational climate and leadership style affects performance. *Management Decision*, *46*(3), 406–432. doi:10.1108/00251740810863861

Hair, J. F., Black, W. C., Babin, B. J., & Anderson, R. E. (2010). *Multivariate data analysis, a global perspective* (7th ed.). Prentice-Hall.

Hallinger, P., & Kovačević, J. (2019). Science mapping the knowledge base in educational leadership and management: A longitudinal bibliometric analysis, 1960 to 2018. *Educational Management Administration & Leadership*, *49*(1), 5–30. doi:10.1177/1741143219859002

Hall, R. D., & Rowland, C. A. (2016). Leadership development for managers in turbulent times. *Journal of Management Development*, *35*(8), 942–955. doi:10.1108/JMD-09-2015-0121

Hamilton, C. A. (2007). Have a High copper Dissolution Rates of SAC305/405 Alloys Forced. *SMTAi, PanPacific Microelectronics Symposium.*

Hasan, S., Ali, M., Kurnia, S., & Thurasamy, R. (2021). Evaluating the cyber security readiness of organizations and its influence on performance. *Journal of Information Security and Applications*, *102726*, 102726. doi:10.1016/j.jisa.2020.102726

Hernandez, J. G., & Salcedo, M. T. (2020). The influence and benefits of technology as a strategy in organizations. *Multidisciplinary Journal for Education Social and Technological Sciences*, *7*(1), 32–53. doi:10.4995/muse.2020.10693

Héroux, S., & Fortin, A. (2022). Board of directors' attributes and aspects of cybersecurity disclosure. *The Journal of Management and Governance*. doi:10.100710997-022-09660-7

Herwan, J., Kano, S., Oleg, R., Sawada, H., & Kasashima, N. (2018). *Cyber-Physical System Architecture for Machining Production Line.*

Homoliak, I., Toffalini, F., Guarnizo, J., Elovici, Y., & Ochoa, M. (2020). Insight into insiders and IT: A Survey of insider threat taxonomies, analysis, modeling, and countermeasures. *ACM Computing Surveys*, *52*(30), 1–40. doi:10.1145/3303771

Horn, G., Skrzypek, P., Materka, K., & Przeździęk, T. (2019). Cost benefits of multi-cloud deployment of dynamic computational intelligence applications. In *Workshops of the International Conference on Advanced Information Networking and Applications*, (pp. 1041-1054). Springer. 10.1007/978-3-030-15035-8_102

Hunt, B. W. (2008). Reduce Cupper Disolution on Lead-Free Assembly. National Physic Laboratory.

Hunt, D. D. (2009). On the factors affecting the dissolution of copper in molten lead free solders and development of a method to asses the soldering parameters. Emerald Group Publishing Limited.

Hu, W., Chang, C., Sengupta, A., Bhunia, S., Kastner, R., & Li, H. (2021). An overview of hardware security and trust: Threats, countermeasures, and design tools. *IEEE Transactions on Computer-Aided Design of Integrated Circuits and Systems*, *40*(6), 1010–1038. doi:10.1109/TCAD.2020.3047976

Ibarra, J., & García, F. (2016). Las demandas de factores productivos en la industria maquiladora. *Estudios Económicos*, 265-303.

Institut de technologie de Cambodge, Institute of Electrical and Electronics Engineers. Islamabad Section, & Institute of Electrical and Electronics Engineers. (n.d.). *SKIMA 2018 : the 12th International Conference on Software, Knowledge, Information Management & Applications*. IEEE.

Institute of Electrical and Electronics Engineers. (n.d.). *2018 International Conference on Computing, Electronic and Electrical Engineering (ICE Cube)*. IEEE.

Instituto Nacional de Estadística y Geografía. (2015). *Síntesis metodológica de la estadística del programa de la industria.* INEGI. https://www.inegi.org.mx/est/contenidos/proyectos/accesomicrodatos/immex/default.aspxInstituto Nacional de Geografía y Estadística. (2022). *Manufacturas.* INEGI. https://www.inegi.org.mx/temas/manufacturas/

Iqbal, M. A., Hussain, S., Xing, H., & Imran, M. A. (2020). *Enabling the internet of things: fundamentals, design and applications.* John Wiley & Sons.

Jabbar, M. A. (2023). Applied Machine Learning and Data Analytics. *5th International Conference, AMLDA 2022*. Springer Nature.

Jabbar, M. A. Tiwari, S., Ortiz-Rodriguez, F. (2023). Smart Urban Computing Applications. River Publishers.

Jabbar, M. A., Tiwari, S., & Ortiz-Rodriguez, F. (2023). Smart Urban Computing Applications. River Publishers

Jabbar, M. A., Tiwari, S., & Ortiz-Rodriguez, F. (2023). Smart Urban Computing Applications. River Publishers.

Jafar, M. T., Al-Fawa'reh, M., Barhoush, M., & Alshira'H, M. H. (2022). Enhanced Analysis Approach to Detect Pishing Atacks During COVID-19 Crisis. *Cybernetics and Information Technologies*, *22*(1), 60–76. doi:10.2478/cait-2022-0004

Jagatheesaperumal, S. K., & Rahouti, M. (2022). Building Digital Twins of Cyber Physical Systems With Metaverse for Industry 5.0 and Beyond. *IT Professional*, *24*(6), 34–40. doi:10.1109/MITP.2022.3225064

Jarrahi, M. H. (2018). Artificial intelligence and the future of work: human-AI symbiosis in organizational decision making. *Business Horizons*, *61*(4), 577–586. doi:10.1016/j.bushor.2018.03.007

Jeyaraj, A., & Zadeh, A. (2020). Institutional isomorphism in organizational cybersecurity. *Journal of Organizational Computing and Electronic Commerce*, *20*(4), 361–380. doi:10.1080/10919392.2020.1776033

Jiménez, F. (2011). *Crecimiento Económico: Enfoques y Modelos.* Lima, Perú: Fondo editorial de la Pontificia Universidad Católica del PErú.

Jiménez-Valverde, G., & Núñez-Cruz, E. (2009). Online cooperation in virtual environments in the teaching of chemistry. *Chemical Educator (Boise, Idaho)*, *20*(3), 314–319. doi:10.1016/S0187-893X(18)30031-4

Johansen, A. G. (2020). *What is RFID and why does it matter?* Norton. https://us.norton.com/blog/id-theft/what-is- rfid

Kakatkar, C., Bilgram, V., & Füller, J. (2020). Innovation analytics: Leveraging artificial intelligence in the innovation process. *Business Horizons*, *63*(2), 171–181. doi:10.1016/j.bushor.2019.10.006

Kaplan, R. S., & Norton, D. P. (2004). The strategy map: Guide to aligning intangible assets. *Strategy and Leadership*, *32*(5), 10–17. doi:10.1108/10878570410699825

Kim, N., & Lee, S. (2021). Cybersecurity breach and crisis response: An analysis of organizations' official statements in the United States and South Korea. *International Journal of Business Communication*, *58*(4), 560–581. doi:10.1177/2329488418777037

Kim, S., & Park, S. (2017). CPS(Cyber Physical System) based Manufacturing System Optimization. *Procedia Computer Science*, *122*, 518–524. doi:10.1016/j.procs.2017.11.401

Kipper, L. M., Furstenau, L. B., Hoppe, D., Frozza, R., & Lespen, S. (2019). Scopus scientific mapping production in industry 4.0 (2011-2018): A bibliometric analysis. *International Journal of Production Research*, 1–24. doi:10.1080/00207543.2019.1671625

Kraaijenbrink, J. (2022). What is Industry 5.0 and how it will change companies. *Forbes*. https://forbes.es/empresas/167359/que-es-la-industria-5-0-y-como-cambiara-las-empresas/

Krol, M., Brouwer, W., & Rutten, F. (2013). Productivity costs in economic evaluations: Past, present, future. *PharmacoEconomics*, *31*(7), 537–549. doi:10.100740273-013-0056-3 PMID:23620213

Kure, H. I., Islam, S., & Razzaque, M. A. (2018). An Integrated Cyber Security Risk Management Approach for a Cyber-Physical System. *Applied Sciences (Basel, Switzerland)*, *8*(6), 898. doi:10.3390/app8060898

Lampe, B. T. (1976). Room Temperature Aging propierties of some solder alloys. *Welding Journal*, 330s–340s.

Langle, M., Méndez, O., & Sánchez, J. (2021). Factores predictores del índice de rotación de personal: el caso de una empresa maquiladore en Reynosa. *Análisis Económico*, 119-140.

Laplante, P. A. (2000). *Dictionary of computer science, Engineering and Technology*. CRC Press.

Latilla, V. M., Urbinati, A., Cavallo, A., Franzo, S., & Ghezzi, A. (2021). Organizational re-design for business model innovation while exploiting digital technologies: A single case study of an energy company. *International Journal on Management of Innovation & Technology*, *18*(2), 2040002. doi:10.1142/S0219877020400027

Leau, Y. B., Loo, W. K., Tham, W. Y., & Tan, S. F. (2012). Software development life cycle AGILE vs traditional approaches. *Int Conf Inf Netw Technol.*, *37*(1), 162–167.

Lee, S., Abdullah, A., Jhanjhi, N., & Kok, S. (2021). Classification of botnet attacks in IoT smart factory using honeypot combined with machine learning. *PeerJ. Computer Science*, *7*, e350. doi:10.7717/peerj-cs.350 PMID:33817000

Leng, J., Ye, S., Zhou, M., Zhao, J. L., Liu, Q., Guo, W., Cao, W., & Fu, L. (2020). Blockchain-Secured Smart Manufacturing in Industry 4.0: A Survey. *IEEE Transactions on Systems, Man, and Cybernetics. Systems*, *51*(1), 237–252. doi:10.1109/TSMC.2020.3040789

Levin, R. I. (2006). Capitulo 7. Muestreo y distribuciones de muestreo. In R. I. Berenso (Ed.), *Estadistica para la administracion (págs* (pp. 206–214). Pearson.

Lidong, W., & Guanghui, W. (2016). Big Data in Cyber-Physical Systems, Digital Manufacturing and Industry 4.0. *International Journal of Engineering and Manufacturing*, *6*(4), 1–8. doi:10.5815/ijem.2016.04.01

Li, L., He, W., Xu, L., Ash, I., Anwar, M., & Yuan, X. (2019). Investigating the impact of cybersecurity policy awareness on employees' cybersecurity behavior. *International Journal of Information Management*, *45*, 13–24. doi:10.1016/j.ijinfomgt.2018.10.017

Linares, M. (2012). *Cybercrime Phenomenology and criminology of crime in cyberspace*. Marcial Pons. https://soyancrig.com.gt/data/files/libros/el-cibercrimen.pdf

Liu, H., & Wang, L. (2020). Remote human–robot collaboration: A cyber–physical system application for hazard manufacturing environment. *Journal of Manufacturing Systems*, *54*, 24–34. doi:10.1016/j.jmsy.2019.11.001

Loan, F. A., Bisma, B., & Nahida, N. (2022). Global research productivity in cybersecurity: A scientometric study. *Global Knowledge. Memory and Communication*, *7*(4/5), 342–354. doi:10.1108/GKMC-09-2020-0148

Loayza, N. V., & Raddatz, C. (2010). The composition of growth matters for poverty alleviation. *Journal of Development Economics*, *93*(1), 137–151. doi:10.1016/j.jdeveco.2009.03.008

Lopez, S. I. M., & Valenzuela, B. G. E. (2015). CHILDREN and adolescents with special educational needs. *Clínica Las Condes Medical Journal*, *26*(1), 42–51. doi:10.1016/j.rmclc.2015.02.004

López-Vázquez, F., Sahagún, M. A., & Ortiz-Rodriguez, F. (2017). Análisis y comprensión de factores que influyen en la productividad empresarial. *Congreso Internacional Multidiciplinario Veracruz 2017 Tecnologico Nacional de Mexico*. Instituto Tecnologico de Veracruz. Revista Impulso Tecnológico.

López-Vázquez,. F. & Sahagún, M. A. (2016). Estudio de Caso: Medicion de Habilidades Humanas y Clima Organizacional para Identificar y Desarrollar Acciones de Productividad Incremental. *20th Annual Western Hemispheric Trade Conference, Laredo, TX, USA*. Free Trade. http://freetrade.tamiu.edu/PDF/20Conf-Sessions.pdf#page=1

López-Vázquez, F., & Sahagún, M. A. (2017). *Análisis y Comprensión de la Relación Liderazgo Productividad en la Industria de Manufactura México-USA: Teoría y Prueba Empírica. II Congreso de Estudios de la frontera*, Brownsville, TX, USA.

López-Vázquez, F., Sahagún, M. A., & Cisneros-Lascurain, E. E. (2017). *Análisis de los principales factores de influencia en el desarrollo de productividad incremental: sector de manufactura electrónica y automotriz. 21th Annual Western Hemispheric Trade Conference*, Laredo, TX, USA.

López-Vázquez, F., Sahagún, M. A., & Ortiz-Rodriguez, F. (2018). *Efectos entre Inequidad Salarial y Productividad de la Industria de Manufactura en la frontera México-Estados Unidos. 22th Annual Western Hemispheric Trade Conference*, Laredo, TX, USA.

Lota, S. (2023). *A Comprehensive Guide to IoT Security*. Nozomi Network. https://www.nozominetworks.com/blog/guide-to-iot-security/

Luca, D., & Dolga, V.-I. (2021). From Industry 3.0 to Industry 4.0. *The International Journal of Engineering and Science*, *10*(9), 44–49.

Mageswaran Arunasalam, Z. B. (2021). Challenges in Minimizing Copper Dissolution for Lead Free Wave in surface mount technology going towards green manufacturing. *International Journal of Precision Engineering and Manufacturing-Green Technology*.

Mahmoud, M. S., & Mohamad, A. A. H. (2016). *A study of efficient power consumption wireless communication techniques/modules for internet of things (IoT) applications*. Research Gate.

Manko, H. I. (1992). *Solders and Soldering*. Mc Graw Hill.

Martinez, R. (2023, May 26). *General Director of Agencia Aduanal Asesoría y Soluciones Aduaneras*. S.C. (P. Ramirez, Interviewer) Reynosa, Tamaulipas, Mexico.

Martínez, S., Flamand, L., & Hernández, A. (2021). Creimiento y desarrollo en los municipios frotnerizos del norte de México: Índice de Desarrollo Municipal Básico y PIB. In *COLEF, 25 años de integración económica en la frontera norte de México (pág. 411)*. Colegio de la Frontera Norte.

Martinic, S., & Villalta, M. (2015). Classroom time management and school performance in full-time schools in Chile. *Educational Profiles*, *37*(147), 28–49. doi:10.1016/j.pe.2013.03.001

Marúm-Espinosa, E., & Reynoso-Cantú, E.-L. (2014). The importance of non-formal education for sustainable human development in Mexico. *Ibero-American Journal of Higher Education*, *5*(12), 137–155. doi:10.1016/S2007-2872(14)71947-X

Massalski, T. (1996). *Binary Alloy Phase Diagrams.* Ohio: 2da Edicion.

Matheson, E., Minto, R., Zampieri, E. G. G., Faccio, M., & Rosati, G. (2019). Human–robot collaboration in manufacturing applications: A review. *Robotics (Basel, Switzerland)*, *8*(4), 100. doi:10.3390/robotics8040100

Mayhew, D. J. (1999). The usability engineering lifecycle. In CHI'99 Extended Abstracts on Human Factors in Computing Systems (pp. 147–8). ACM. doi:10.1145/632716.632805

Mc Carty, M., Johnson, J., Richardson, B., Rieger, C., Cooley, R., Gentle, J., Rothwell, B., Phillips, T., Novak, B., Culler, M., & Wright, B. (2023). Cybersecurity Resilience Demonstration for Wind Energy Sites in Co-Simulation Environment. *IEEE Access : Practical Innovations, Open Solutions*, *11*, 15297–15313. doi:10.1109/ACCESS.2023.3244778

McClelland, D. C., & Boyatzis, R. E. (1982). Leadership motive pattern and long-term success in management. *The Journal of Applied Psychology*, *67*(6), 737–743. doi:10.1037/0021-9010.67.6.737

Medina-Quintero, J. M., Sahagun, M. A., Alfaro, J., & Ortiz-Rodriguez, F. (2023). *Global Perspectives on the Strategic Role of Marketing Information Systems*. doi:10.4018/978-1-6684-6591-2

Meinke, K., & Bennaceur, A. (2018). Machine learning for software engineering: models, methods, and applications. In *2018 IEEE/ACM 40th International Conference on Software Engineering: Companion (ICSE-Companion).* IEEE. 10.1145/3183440.3183461

Meixell, M. J., & Gargeya, V. B. (2005). Global supply chain design: A literature review and critique. *Transportation Research Part E, Logistics and Transportation Review*, *41*(6), 531–550. doi:10.1016/j.tre.2005.06.003

Mendoza, J. E.-P. (2022). Comercio exterior y empleo en la frontera norte de México. (U. A. California, Ed.) Estudios Fronterizos, 23. doi:10.21670/ref.2207091

Mendoza-Rojas, J. (2015). Expansion of the offer of higher education in Mexico and creation of public institutions 2001-2012. *Ibero-American Journal of Higher Education*, *6*(16), 3–32. doi:10.1016/j.rides.2015.04.001

Meneghello, F., Calore, M., Zucchetto, D., Polese, M., & Zanella, A. (2019). IoT: Internet of Threats? A Survey of Practical Security Vulnerabilities in Real IoT Devices. *IEEE Internet of Things Journal*, *6*(5), 8182–8201. doi:10.1109/JIOT.2019.2935189

Michalakopoulou, K., Nikita, A., Njoya, E. T., & Johnes, J. (2022). Innovation in the legal service industry: Examining the roles of human and social capital, and knowledge and technology transfer. *International Journal of Entrepreneurship and Innovation*. doi:10.1177/14657503221119667

Miller, A. (2019). Permissioned and permissionless blockchains. *Blockchain for Distributed Systems Security*, 193–204.

MIT Technology Review. (2023). The 10 Emerging Technologies. *MIT Technology Review*. https://www.technologyreview.es//listas/tecnologias-emergentes/2023

Moeuf, A., Pellerin, R., Lamouri, S., Tamayo-Giraldo, S., & Barbaray, R. (2018). The industrial management of SMEs in the era of Industry 4.0. *International Journal of Production Research*, *56*(3), 1118–1136. doi:10.1080/00207543.2017.1372647

Montalvo, J. G., & Ravallion, M. (2010). The pattern of growth and poverty reduction in China. *Journal of Comparative Economics*, *38*(1), 2–16. doi:10.1016/j.jce.2009.10.005

Moore, T. (2010). The economics of cybersecurity: Principles and policy options. *International Journal of Critical Infrastructure Protection*, *3*(3-4), 103–117. doi:10.1016/j.ijcip.2010.10.002

Morgan, S. (2020). Cybercrime to cost the world $10.5 trillion annually by 2025. *Cybercrime Magazine*. https://cybersecurityventures.com/hackerpocalypse-cybercrime-report-2016/

Mueller, M. L. (2020). Against sovereignty in cyberspace. *International Studies Review*, *22*(4), 779–801. doi:10.1093/isr/viz044

Muenchaisri P. (2019). *Literature reviews on applying artificial intelligence/machine learning to software engineering research problems: preliminary*. Academic Press.

Nasereldin, Y. A., Brenya, R., Bassey, A. P., Ibrahim, I. E., Alnadari, F., Nasiru, M. M., & Ji, Y. (2020). Is the global food supply chain during the COVID-19 pandemic resilient? A review paper. *Open Journal of Business and Management*, *9*(1), 184–195. doi:10.4236/ojbm.2021.91010

Ng, I. C., & Wakenshaw, S. Y. (2017). The Internet-of-Things: Review and research directions. *International Journal of Research in Marketing*, *34*(1), 3–21. doi:10.1016/j.ijresmar.2016.11.003

Nicol, H. N., & Minghi, J. (2005). The continuing relevance of borders in contemporary contexts. *Geopolitics*, *10*(4), 680–687. doi:10.1080/14650040500436647

Nikolakis, N., Alexopoulos, K., Xanthakis, E., & Chryssolouris, G. (2019). The digital twin implementation for linking the virtual representation of human-based production tasks to their physical counterpart in the factory-floor. *International Journal of Computer Integrated Manufacturing*, *32*(1), 1–12. doi:10.1080/0951192X.2018.1529430

Núñez-Peña, M., Bono, R., & Suárez-Pellicioni, M. (2015). Formative Assessment in Higher Education: Impact on Students with Math Anxiety. *Procedia - Behavioral and Social Sciences*, *196* (July 2014), 135-141. doi:10.1016/j.sbspro.2015.07.023

O'Connell, E., Moore, D., & Newe, T. (2020). Challenges associated with implementing 5G in manufacturing. *Telecom*, *1*(1), 5. doi:10.3390/telecom1010005

Office of the European Union. (2022). *Industry 5.0: Towards a sustainable, human-centric and resilient European industry*. Europea. https://op.europa.eu/en/publication-detail/-/publication/468a892a-5097-11eb-b59f-01aa75ed71a1/

Oliveto, R., Antoniol, G., Marcus, A., & Hayes, J. (2007). Software Artefact Traceability: The Never-Ending Challenge. In *2007 IEEE International Conference on Software Maintenance* (pp. 485–488). IEEE. 10.1109/ICSM.2007.4362664

Olsen, J. A., & Richardson, J. (1999). Production gains from health care: What should be included in cost-effectiveness analyses? *Social Science & Medicine*, *49*(1), 17–26. doi:10.1016/S0277-9536(99)00116-1 PMID:10414837

Onumo, A., Ullah-Awan, I., & Cullen, A. (2021). Assessing the moderating effect of security technologies on employees compliance with cybersecurity control procedures. [TMIS]. *ACM Transactions on Management Information Systems*, *12*(2), 1–29. doi:10.1145/3424282

Orozco, J. E. F., Becerra, J. I. V., & Arellano, C. I. M. (2015). Current Perspectives on World University Rankings. *The Journal of Higher Education*, *44*(175), 41–67. doi:10.1016/j.resu.2015.09.001

Ortiz-Rodriguez, F., Barrera, R., Villanueva, V., Medina, M., & Gauna, A. (2020). APPVISO: citizen digital communication to prevent risk events in real time. In *The 21st Annual International Conference on Digital Government Research (dg.o '20)*. Association for Computing Machinery. 10.1145/3396956.3397019

Ortiz-Rodriguez, F., Tiwari, S., Sicilia, M. A., & Nikiforova, A. (2023). *Electronic Governance with Emerging Technologies*. First International Conference, EGETC 2022, Tampico, Mexico. https://www.konradlorenz.edu.co/blog/tecnologias-emergentes-y-su-impacto-en-las-organizaciones/

Ortiz-Rodriguez, F., Tiwari, S., Sicilia, M. A., & Nikiforova, A. (2023). *Electronic Governance with Emerging Technologies*. First International Conference, EGETC 2022, Tampico, Mexico. Revised Selected Papers.

Ortiz-Rodriguez, F., Tiwari, S., Sicilia, M., & Nikiforova, A. (2023). *Electronic Governance with Emerging Technologies*. First International Conference, EGETC 2022, Tampico, Mexico.

Ortiz-Rodriguez, F., Tiwari, S., Sicilia, M. A., & Nikiforova, A. (2023). Electronic Governance with Emerging Technologies. *First International Conference, EGETC 2022*, Tampico, Mexico.

Oseguera, A. (2022). Concentración en el mercado laboral y su relación con los salarios en México. *Estudios Económicos*, 45-102.

Padmanaban, P.H., & Sharma, Y.K. (2019). Implication of artificial intelligence in software development life cycle: a state-of-the-art review. *IJRRA*.

Pan, Y., & Jackson, R. T. (2008). Ethnic difference in the relationship between acute inflammation and serum ferritin in US adult males. *Epidemiology and Infection*, *136*(3), 421–431. doi:10.1017/S095026880700831X PMID:17376255

Papakostas, N., Newell, A., & Hargaden, V. (2019). A novel paradigm for managing the product development process utilising blockchain technology principles. *CIRP Annals*, *68*(1), 137–140. doi:10.1016/j.cirp.2019.04.039

Parkin, M., & Loría, E. (2010). *Microeconomía. Versión para Latinoamérica*. PEARSON EDUCACIÓN.

Park, Y., El Sawy, O. A., & Fiss, P. C. (2017). The role of business intelligence and communication technologies in organizational agility: A configurational approach. *Journal of the Association for Information Systems*, *18*(9), 648–686. doi:10.17705/1jais.00467

Patel, S., Graham, J., & Ralston, P. (2008). Quantitatively assessing the vulnerability of critical information systems: A new method for evaluating security enhancements. *International Journal of Information Management*, *28*(6), 483–491. doi:10.1016/j.ijinfomgt.2008.01.009

Pathak, P., Pal, P. R., Shrivastava, M., & Ora, P. (2019). Fifth revolution: Applied AI & human intelligence with cyber physical systems. *International Journal of Engineering and Advanced Technology*, *8*(3), 23–27.

Philip, S. J., Luu, T., & Carte, T. (2023). There's no place like home: Understanding users' intentions toward securing Internet-of-thing (IoT) smart home networks. *Computers in Human Behavior*, *107551*(139), 107551. doi:10.1016/j.chb.2022.107551

Pivoto, D. G. S., de Almeida, L. F. F., da Rosa Righi, R., Rodrigues, J. J. P. C., Lugli, A. B., & Alberti, A. M. (2021). Cyber-physical systems architectures for industrial internet of things applications in Industry 4.0: A literature review. *Journal of Manufacturing Systems*, *58*, 176–192. doi:10.1016/j.jmsy.2020.11.017

Poole, D. L., & Mackworth, A. K. (2010). *Artificial intelligence: foundations of computational agents*. Cambridge University Press. doi:10.1017/CBO9780511794797

Pranggono, B., & Arabo, A. (2020). COVID-19 Pandemic cybersecurity issues. *Internet Technology Letters*, *4*(2). doi:10.1002/itl2.247

Quintero, M., Melchor, J., & Echeverria, R. Osiris & Ortiz-Rodriguez, F. (2021). *Confianza y calidad de la información para la satisfacción y lealtad del cliente en el e-Banking con el uso del teléfono celular.* Contaduría y Administración. doi:10.22201/fca.24488410e.2022.2965

Radanliev, P., De Roure, D., Van Kleek, M., Santos, O., & Ani, U. (2021). Artificial intelligence in cyber physical systems. *AI & Society*, *36*(3), 783–796. doi:10.100700146-020-01049-0 PMID:32874020

Raja, A. S., & Palanichamy, P. (2011). Leadership Styles and Its Impact on Organizational Commitment. *The Journal of Commerce, 3*(4). Hailey College of Commerce, University of the Punjab, Pakistan.

Ramadan, R., & Widyani, Y. (2013). Game development life cycle guidelines. In *2013 International Conference on Advanced Computer Science and Information Systems (ICACSIS)*. IEEE.

Rao, S. K., & Prasad, R. (2018). Impact of 5G technologies on industry 4.0. *Wireless Personal Communications*, *100*(1), 145–159. doi:10.100711277-018-5615-7

Rappe, C., & Zwick, T. (2006). Developing leadership competence of production unit managers. *Journal of Management Development*, *26*(4), 312–330. doi:10.1108/02621710710740084

Ravallion, M., & Datt, G. (2002). Why has economic growth been more pro-poor in some states of India than others? *Journal of Development Economics*, *68*(2), 381–400. doi:10.1016/S0304-3878(02)00018-4

Rees, L. P., Deame, J. K., Rakes, T. R., & Baker, W. H. (2011). Decision support for cybersecurity risk planning. *Decision Support Systems*, *51*(3), 493–505. doi:10.1016/j.dss.2011.02.013

Reynosa, M. (2021). *Plan Municipal de Desarrollo 2021-2024*. Municipo de Reynosa Tamaulipas.

Rodríguez, G., Ariza, M., & Ramos, J. L. (2014). Institutional quality and academic performance The case of the Colombian Caribbean universities. *Educational Profiles*, *36*(143), 10–29. doi:10.1016/S0185-2698(14)70607-5

Rogerson, P. A. (2001). *Statistical methods for geography*. Sage. doi:10.4135/9781849209953

Rojas, MS (2014). Violence against women in the news in Mexico: the uncritical and uneven treatment of a serious social problem1. *Sociological Act, 65*(2012), 37-64. . doi:10.1016/S0186-6028(14)70236-1

Ruparelia, N. B. (2010). Software development lifecycle models. *Software Engineering Notes*, *35*(3), 8–13. doi:10.1145/1764810.1764814

Sánchez Juárez, I. L., & Campos Benítez, E. (2010). Industria manufacturera y crecimiento económico en la frontera norte de México. *Región y Sociedad*, *22*(49), 45–89.

Sánchez, R. B. (2021). *Home*. Diario Oficial de la Federacion. https://www.dof.gob.mx/nota_detalle.php?codigo=5639315&fecha=24/12/2021#gsc.tab=0

Sanchez, R. (2013). Reflecting on the Border. A Search for Alternative Research Approaches. EL colegio de la Frontera. *Frontera Norte*, *26*(3).

Satyanarayan, K. P. (2002). *Wetting Behavior of Lead-Free Solders On*. Department of Metallurgical & Materials Engineering National Institute of Technology Karnataka Surathkal,

Savchenko, D., Kasurinen, J., & Taipale, O. (2019). Smart tools in software engineering: a systematic mapping study. In *2019 42nd International Convention on Information and Communication Technology, Electronics and Microelectronics (MIPRO)*. IEEE. 10.23919/MIPRO.2019.8756975

Schuck, A. M., & Rabe-Hemp, C. E. (2018). Investing in people: Salary and turnover in policing. *Policing*, *41*(1), 113–128. doi:10.1108/PIJPSM-09-2016-0137

Secretaria de Economía. (2010). *Secretaria de Economía.* Economia. http://www.2006-2012.economia.gob.mx/comunidad-negocios/industria-y-comercio/instrumentos-de-comercio-exterior/immex

Seffah, A., Gulliksen, J., & Desmarais, M. C. (2005). Human-centered software engineering-integrating usability in the software development lifecycle (Vol. 8). Luxemburg: Springer Science & Business Media.

Servico Nacional de Informacion de Comercio Exterior. (n.d.). *IMMEX: Acerca de.* Servicio Nacional de Informacion de Comercio Exterior. https://www.snice.gob.mx/cs/avi/snice/progfom.immex.acercade.html#:~:text=El%20Programa%20IMMEX%2C%20es%20un,de%20servicios%20de%20exportaci%C3%B3n%2C%20difiriendo

Sharma, A., & Singh, B. J. (2020). Evolution of industrial revolutions: A review. *International Journal of Innovative Technology and Exploring Engineering*, *9*(11), 66–73. doi:10.35940/ijitee.I7144.0991120

Shetty, S. H., Rao, A., & Gatti, R. R. (2021). *State of the Art Review of IIoT Communication Protocols.*, doi:10.4018/978-1-7998-7541-3.ch003

Sheu, H. J., & Yang, C. Y. (2005). Insider Ownership Structure and Firm Performance: A productivity perspective study in Taiwan's electronics industry. *Corporate Governance*, *13*(2), 326–337. doi:10.1111/j.1467-8683.2005.00426.x

Showalter, C., Sibell, S., Jean, D., Klimah, P., & Biggs, J. (2016). A Study of Overcoming Solder Icicling And Copper Wire Dissolution. In *An Automated Lead-Free Soldering System.* Kester Inc.

Singhal, B., Dhameja, G., & Panda, P. S. (2018). How blockchain works. In *Beginning blockchain* (pp. 31–148). Springer. doi:10.1007/978-1-4842-3444-0_2

Sirmon, D. G., Hitt, M. A., & Ireland, R. D. (2007). Managing firm resources in dynamic environments to create value: Looking inside the black box. *Academy of Management Review*, *32*(1), 273–292. doi:10.5465/amr.2007.23466005

Sittig, D. F., & Singh, H. (2016). A socio-technical approach to preventing, mitigating, and recovering from ransomware attacks. *Applied Clinical Informatics*, *7*(2), 624–632. doi:10.4338/ACI-2016-04-SOA-0064 PMID:27437066

Smith, T., Tadesse, A., & Vincent, N. E. (2021). The impact of CIO characteristics on data breaches. *International Journal of Accounting Information Systems*, *100532*, 100532. doi:10.1016/j.accinf.2021.100532

Snyder, H. (2019). Literature review as a research methodology: An overview and guidelines. *Journal of Business Research*, *104*, 333–339. doi:10.1016/j.jbusres.2019.07.039

Solomon, J. (2003) *Who's Counting? A Lean Accounting Business Novel.* WCM Associates.

Sorte, B.W., Joshi, P.P., & Jagtap, V. (2015). Use of artificial intelligence in software development life cycle—a state of the art review. *International Journal of Advanced Engineering and Global Technology*, 398–403.

Spajke, R. B. (2013). *Dissolution of a Copper Wire During a Hot Dipping Process Using Sncul Lead Free Solder*. Institute of Metals and Technology.

Sparrow, A. (2022). *Leveraging erp for smarter manufacturing (MOM)*. LinkedIn. https://www.linkedin.com/pulse/leveraging-erp-smarter-manufacturing-mom-andrew-sparrow

Spear, B. (2008). James Watt: The steam engine and the commercialization of patents. *World Patent Information*, *30*(1), 53–58. doi:10.1016/j.wpi.2007.05.009

Stetsenko, S., Tytok, V., Emelianova, O., Bielienkova, O., & Tsyfra, T. (2021). The Interrelation of Digital Technologies and Organizational and Economic Mechanisms in Construction: Adaptation to Change Management. *International Review (Steubenville, Ohio)*, *1-2*(1), 21–31.

Stylianou, C., & Andreou, A. S. (2016). Investigating the impact of developer productivity, task interdependence type and communication overhead in a multi-objective optimization approach for software project planning. *Advances in Engineering Software*, *98*, 79–96. doi:10.1016/j.advengsoft.2016.04.001

Sukumar, A., Mahdiraji, H. A., & Jarari-Sadeghi, V. (2023). Cyber risk assessment in small and medium-sized enterprises: A multi-level decision-making approach for small e-tailors. *Risk Analysis*, risa.14092. doi:10.1111/risa.14092 PMID:36627823

Sun, N., Zhang, J., Rimba, P., Gao, S., Zhang, L. Y., & Xiang, Y. (2018). Data-Driven Cybersecurity Incident Prediction: A Survey. *IEEE Communications Surveys and Tutorials*, *21*(2), 1744–1772. doi:10.1109/COMST.2018.2885561

Tanabe, G. I. (2007). *Dissolution of copper on Sn-Ag-Cu system lead*. Emerald Group.

Tantawi, K. H., Sokolov, A., & Tantawi, O. (2019). Advances in industrial robotics: From industry 3.0 automation to industry 4.0 collaboration. *2019 4th Technology Innovation Management and Engineering Science International Conference (TIMES-ICON)*, (pp. 1–4). IEEE.

Tao, F., Qi, Q., Wang, L., & Nee, A. Y. C. (2019). Digital twins and cyber–physical systems toward smart manufacturing and industry 4.0: Correlation and comparison. *Engineering (Beijing)*, *5*(4), 653–661. doi:10.1016/j.eng.2019.01.014

Tarallo, A., Mozzillo, R., Di Gironimo, G., & De Amicis, R. (2018). A cyber-physical system for production monitoring of manual manufacturing processes. *International Journal on Interactive Design and Manufacturing*, *12*(4), 1235–1241. doi:10.100712008-018-0493-5

Tetsuro, N. &. (2001). *A control Method for Copper Content in a Solder Dipping Bath*. World Intellectual property organization.

Tobón, O. E. A., Zapata, S. J. C., Lopera, I. C. P., & Duque, J. W. S. (2014). Academic training, values, empathy and socially responsible behaviors in university students. *The Journal of Higher Education*, *43*(169), 89–105. doi:10.1016/j.resu.2015.01.003

Tributaria, S. D. (2017). *Home.* SAT. https://www.sat.gob.mx/consulta/78455/exportadores

Tributaria, S. D. (2018). *Home.* SAT. https://www.sat.gob.mx/consulta/54336/importadores

Triola, M. F. (2009). *Estadistica, Decima Edicion.* Pearson,Addison Wesley.

Tucker, R. S. (2010). Green optical communications—Part I: Energy limitations in transport. *IEEE Journal of Selected Topics in Quantum Electronics*, *17*(2), 245–260. doi:10.1109/JSTQE.2010.2051216

Ullah, F., Naeem, H., Jabbar, S., Khalid, S., Latif, M., Al-Turjam, F., & Mostarda, L. (2019). Cyber security threats detection in internet of things using deep learning approach. *IEEE Access : Practical Innovations, Open Solutions*, *7*, 124379–124389. doi:10.1109/ACCESS.2019.2937347

Vaculoy, Alpha Solder. (2022). *Technical Reference Data Sheet for Solder SAC 305.* North America: Mermaid Trademark.

Van Eck, N. J., & Waltman, L. (2010). Software survey: VOSviewer, a computer program for bibliometric mapping. *Scientometrics*, *84*(2), 523–538. doi:10.100711192-009-0146-3 PMID:20585380

Van Hoorn, A., Frey, S., Goerigk, W., Hasselbring, W., Knoche, H., Köster, S., & Wittmüss, N. (2011). *Lübeck: Dynamic analysis for model-driven software modernization.* DynaMod project.

Vatankhah Barenji, A., Li, Z., Wang, W. M., Huang, G. Q., & Guerra-Zubiaga, D. A. (2020). Blockchain-based ubiquitous manufacturing: A secure and reliable cyber-physical system. *International Journal of Production Research*, *58*(7), 2200–2221. doi:10.1080/00207543.2019.1680899

Velmourougan, S., Dhavachelvan, P., Baskaran, R., & Ravikumar, B. (2014). Software development life cycle model to build software applications with usability. In *2014 International Conference on Advances in Computing, Communications and Informatics (ICACCI).* IEEE. 10.1109/ICACCI.2014.6968610

Vitronics Soltec Company. (2011). Technical Data for Delta 5, Vitronics. China: Suzhou.

Vrhovec, S., & Mihelic, A. (2021). Redefining threat appraisals of organizational insiders and exploring the moderating role of fear in cyberattack protection motivation. *Computers & Security*, *106*, 102309. doi:10.1016/j.cose.2021.102309

Wang, L., Liu, S., Liu, H., & Wang, X. V. (2020). Overview of human-robot collaboration in manufacturing. *Proceedings of 5th International Conference on the Industry 4.0 Model for Advanced Manufacturing: AMP 2020*, (pp. 15–58). Springer. 10.1007/978-3-030-46212-3_2

Wan, J., Chen, M., Xia, F., Di, L., & Zhou, K. (2013). From machine-to-machine communications towards cyber-physical systems. *Computer Science and Information Systems*, *10*(3), 1105–1128. doi:10.2298/CSIS120326018W

Wan, J., Yan, H., Liu, Q., Zhou, K., Lu, R., & Li, D. (2013). Enabling cyber–physical systems with machine–to–machine technologies. *International Journal of Ad Hoc and Ubiquitous Computing*, *13*(3–4), 187–196. doi:10.1504/IJAHUC.2013.055454

Well, P. C. B. (2007). *2007-2021 WellPCB Technology Co., Ltd.* PlacaPCB. https://placapcb.com/OSP-PCB-Surface-Finish.html

Werner, O., & Campbell, D. T. (1970). Translating, working through interpreters, and the problem of decentering. In R. Naroll & R. Cohen (Eds.), *A handbook of method in cultural anthropology* (pp. 398–420). Natural History Press.

Willis, D. D. (2008). *Measurements of Copper Disolution in lead free solders.* United kingdom: Depratment for Inovation Universities & Skills.

Wooldridge, J. (2013). *Introducción a la Econometría.* Cengage Learning.

Yang, L., Li, P., Zhang, Y., Yang, X., Xiang, Y., & Zhou, W. (2018). Effective Repair Strategy Against Advanced Persistent Threat: A Differential Game Approach. *IEEE Transactions on Information Forensics and Security*, *14*(7), 1713–1728. doi:10.1109/TIFS.2018.2885251

Yen, Y. W., Chou, W. T., Tseng, Y. U., Lee, C., & Hsu, C. L. (2008). Investigation of dissolution behavior of metallic substrates and intermetallic compound in molten lead-free solders. *Journal of Electronic Materials*, *37*(1), 73–83. doi:10.100711664-007-0266-6

Yin, Y., Stecke, K. E., & Li, D. (2018). The evolution of production systems from Industry 2.0 through Industry 4.0. *International Journal of Production Research*, *56*(1–2), 848–861. doi:10.1080/00207543.2017.1403664

Yong, L. A., Rivas, L. A., & Chaparro, J. (2010). Technological acceptance model (TAM): A study of the influence of the national culture and of the user profile in the use of ICTS. *Innovar (Universidad Nacional de Colombia)*, *20*(36), 187–203.

Zhang, H., Feng, M., Long, K., Karagiannidis, G. K., & Nallanathan, A. (2019). Artificial intelligence-based resource allocation in ultradense networks: Applying event-triggered Q-learning algorithms. *IEEE Vehicular Technology Magazine*, *14*(4), 56–63. doi:10.1109/MVT.2019.2938328

Zhang, Y., Robinson, D. K., Porter, A. L., Zhu, D., Zhang, G., & Lu, J. (2016). Technology roadmapping for competitive technical intelligence. *Technological Forecasting and Social Change*, *110*, 175–186. doi:10.1016/j.techfore.2015.11.029

Ziapour, A., Khatony, A., Kianipour, N., & Jafary, F. (2015). Identification and analysis of labor productivity components based on achieve model (case study: Staff of Kermanshah university of medical sciences). *Global Journal of Health Science*, *7*(1), 315. PMID:25560364

Related References

To continue our tradition of advancing information science and technology research, we have compiled a list of recommended IGI Global readings. These references will provide additional information and guidance to further enrich your knowledge and assist you with your own research and future publications.

Abdul Razak, R., & Mansor, N. A. (2021). Instagram Influencers in Social Media-Induced Tourism: Rethinking Tourist Trust Towards Tourism Destination. In M. Dinis, L. Bonixe, S. Lamy, & Z. Breda (Eds.), *Impact of New Media in Tourism* (pp. 135-144). IGI Global. https://doi.org/10.4018/978-1-7998-7095-1.ch009

Abir, T., & Khan, M. Y. (2022). Importance of ICT Advancement and Culture of Adaptation in the Tourism and Hospitality Industry for Developing Countries. In C. Ramos, S. Quinteiro, & A. Gonçalves (Eds.), *ICT as Innovator Between Tourism and Culture* (pp. 30–41). IGI Global. https://doi.org/10.4018/978-1-7998-8165-0.ch003

Abtahi, M. S., Behboudi, L., & Hasanabad, H. M. (2017). Factors Affecting Internet Advertising Adoption in Ad Agencies. *International Journal of Innovation in the Digital Economy*, *8*(4), 18–29. doi:10.4018/IJIDE.2017100102

Afenyo-Agbe, E., & Mensah, I. (2022). Principles, Benefits, and Barriers to Community-Based Tourism: Implications for Management. In I. Mensah & E. Afenyo-Agbe (Eds.), *Prospects and Challenges of Community-Based Tourism and Changing Demographics* (pp. 1–29). IGI Global. doi:10.4018/978-1-7998-7335-8.ch001

Agbo, V. M. (2022). Distributive Justice Issues in Community-Based Tourism. In I. Mensah & E. Afenyo-Agbe (Eds.), *Prospects and Challenges of Community-Based Tourism and Changing Demographics* (pp. 107–129). IGI Global. https://doi.org/10.4018/978-1-7998-7335-8.ch005

Agrawal, S. (2017). The Impact of Emerging Technologies and Social Media on Different Business(es): Marketing and Management. In O. Rishi & A. Sharma (Eds.), *Maximizing Business Performance and Efficiency Through Intelligent Systems* (pp. 37–49). Hershey, PA: IGI Global. doi:10.4018/978-1-5225-2234-8.ch002

Ahmad, A., & Johari, S. (2022). Georgetown as a Gastronomy Tourism Destination: Visitor Awareness Towards Revisit Intention of Nasi Kandar Restaurant. In M. Valeri (Ed.), *New Governance and Management in Touristic Destinations* (pp. 71–83). IGI Global. https://doi.org/10.4018/978-1-6684-3889-3.ch005

Alkhatib, G., & Bayouq, S. T. (2021). A TAM-Based Model of Technological Factors Affecting Use of E-Tourism. *International Journal of Tourism and Hospitality Management in the Digital Age*, *5*(2), 50–67. https://doi.org/10.4018/IJTHMDA.20210701.oa1

Altinay Ozdemir, M. (2021). Virtual Reality (VR) and Augmented Reality (AR) Technologies for Accessibility and Marketing in the Tourism Industry. In C. Eusébio, L. Teixeira, & M. Carneiro (Eds.), *ICT Tools and Applications for Accessible Tourism* (pp. 277-301). IGI Global. https://doi.org/10.4018/978-1-7998-6428-8.ch013

Anantharaman, R. N., Rajeswari, K. S., Angusamy, A., & Kuppusamy, J. (2017). Role of Self-Efficacy and Collective Efficacy as Moderators of Occupational Stress Among Software Development Professionals. *International Journal of Human Capital and Information Technology Professionals*, *8*(2), 45–58. doi:10.4018/IJHCITP.2017040103

Aninze, F., El-Gohary, H., & Hussain, J. (2018). The Role of Microfinance to Empower Women: The Case of Developing Countries. *International Journal of Customer Relationship Marketing and Management*, *9*(1), 54–78. doi:10.4018/IJCRMM.2018010104

Antosova, G., Sabogal-Salamanca, M., & Krizova, E. (2021). Human Capital in Tourism: A Practical Model of Endogenous and Exogenous Territorial Tourism Planning in Bahía Solano, Colombia. In V. Costa, A. Moura, & M. Mira (Eds.), *Handbook of Research on Human Capital and People Management in the Tourism Industry* (pp. 282–302). IGI Global. https://doi.org/10.4018/978-1-7998-4318-4.ch014

Arsenijević, O. M., Orčić, D., & Kastratović, E. (2017). Development of an Optimization Tool for Intangibles in SMEs: A Case Study from Serbia with a Pilot Research in the Prestige by Milka Company. In M. Vemić (Ed.), *Optimal Management Strategies in Small and Medium Enterprises* (pp. 320–347). Hershey, PA: IGI Global. doi:10.4018/978-1-5225-1949-2.ch015

Aryanto, V. D., Wismantoro, Y., & Widyatmoko, K. (2018). Implementing Eco-Innovation by Utilizing the Internet to Enhance Firm's Marketing Performance: Study of Green Batik Small and Medium Enterprises in Indonesia. *International Journal of E-Business Research, 14*(1), 21–36. doi:10.4018/IJEBR.2018010102

Asero, V., & Billi, S. (2022). New Perspective of Networking in the DMO Model. In M. Valeri (Ed.), *New Governance and Management in Touristic Destinations* (pp. 105–118). IGI Global. https://doi.org/10.4018/978-1-6684-3889-3.ch007

Atiku, S. O., & Fields, Z. (2017). Multicultural Orientations for 21st Century Global Leadership. In N. Baporikar (Ed.), *Management Education for Global Leadership* (pp. 28–51). Hershey, PA: IGI Global. doi:10.4018/978-1-5225-1013-0.ch002

Atiku, S. O., & Fields, Z. (2018). Organisational Learning Dimensions and Talent Retention Strategies for the Service Industries. In N. Baporikar (Ed.), *Global Practices in Knowledge Management for Societal and Organizational Development* (pp. 358–381). Hershey, PA: IGI Global. doi:10.4018/978-1-5225-3009-1.ch017

Atsa'am, D. D., & Kuset Bodur, E. (2021). Pattern Mining on How Organizational Tenure Affects the Psychological Capital of Employees Within the Hospitality and Tourism Industry: Linking Employees' Organizational Tenure With PsyCap. *International Journal of Tourism and Hospitality Management in the Digital Age, 5*(2), 17–28. https://doi.org/10.4018/IJTHMDA.2021070102

Ávila, L., & Teixeira, L. (2018). The Main Concepts Behind the Dematerialization of Business Processes. In M. Khosrow-Pour, D.B.A. (Ed.), Encyclopedia of Information Science and Technology, Fourth Edition (pp. 888-898). Hershey, PA: IGI Global. https://doi.org/ doi:10.4018/978-1-5225-2255-3.ch076

Ayorekire, J., Mugizi, F., Obua, J., & Ampaire, G. (2022). Community-Based Tourism and Local People's Perceptions Towards Conservation: The Case of Queen Elizabeth Conservation Area, Uganda. In I. Mensah & E. Afenyo-Agbe (Eds.), *Prospects and Challenges of Community-Based Tourism and Changing Demographics* (pp. 56–82). IGI Global. https://doi.org/10.4018/978-1-7998-7335-8.ch003

Baleiro, R. (2022). Tourist Literature and the Architecture of Travel in Olga Tokarczuk and Patti Smith. In R. Baleiro & R. Pereira (Eds.), *Global Perspectives on Literary Tourism and Film-Induced Tourism* (pp. 202-216). IGI Global. https://doi.org/10.4018/978-1-7998-8262-6.ch011

Barat, S. (2021). Looking at the Future of Medical Tourism in Asia. *International Journal of Tourism and Hospitality Management in the Digital Age, 5*(1), 19–33. https://doi.org/10.4018/IJTHMDA.2021010102

Barbosa, C. A., Magalhães, M., & Nunes, M. R. (2021). Travel Instagramability: A Way of Choosing a Destination? In M. Dinis, L. Bonixe, S. Lamy, & Z. Breda (Eds.), *Impact of New Media in Tourism* (pp. 173-190). IGI Global. https://doi.org/10.4018/978-1-7998-7095-1.ch011

Bari, M. W., & Khan, Q. (2021). Pakistan as a Destination of Religious Tourism. In E. Alaverdov & M. Bari (Eds.), *Global Development of Religious Tourism* (pp. 1-10). IGI Global. https://doi.org/10.4018/978-1-7998-5792-1.ch001

Bartens, Y., Chunpir, H. I., Schulte, F., & Voß, S. (2017). Business/IT Alignment in Two-Sided Markets: A COBIT 5 Analysis for Media Streaming Business Models. In S. De Haes & W. Van Grembergen (Eds.), *Strategic IT Governance and Alignment in Business Settings* (pp. 82–111). Hershey, PA: IGI Global. doi:10.4018/978-1-5225-0861-8.ch004

Bashayreh, A. M. (2018). Organizational Culture and Organizational Performance. In W. Lee & F. Sabetzadeh (Eds.), *Contemporary Knowledge and Systems Science* (pp. 50–69). Hershey, PA: IGI Global. doi:10.4018/978-1-5225-5655-8.ch003

Bechthold, L., Lude, M., & Prügl, R. (2021). Crisis Favors the Prepared Firm: How Organizational Ambidexterity Relates to Perceptions of Organizational Resilience. In A. Zehrer, G. Glowka, K. Schwaiger, & V. Ranacher-Lackner (Eds.), *Resiliency Models and Addressing Future Risks for Family Firms in the Tourism Industry* (pp. 178–205). IGI Global. https://doi.org/10.4018/978-1-7998-7352-5.ch008

Bedford, D. A. (2018). Sustainable Knowledge Management Strategies: Aligning Business Capabilities and Knowledge Management Goals. In N. Baporikar (Ed.), *Global Practices in Knowledge Management for Societal and Organizational Development* (pp. 46–73). Hershey, PA: IGI Global. doi:10.4018/978-1-5225-3009-1.ch003

Bekjanov, D., & Matyusupov, B. (2021). Influence of Innovative Processes in the Competitiveness of Tourist Destination. In J. Soares (Ed.), *Innovation and Entrepreneurial Opportunities in Community Tourism* (pp. 243–263). IGI Global. https://doi.org/10.4018/978-1-7998-4855-4.ch014

Bharwani, S., & Musunuri, D. (2018). Reflection as a Process From Theory to Practice. In M. Khosrow-Pour, D.B.A. (Ed.), Encyclopedia of Information Science and Technology, Fourth Edition (pp. 1529-1539). Hershey, PA: IGI Global. doi:10.4018/978-1-5225-2255-3.ch132

Bhatt, G. D., Wang, Z., & Rodger, J. A. (2017). Information Systems Capabilities and Their Effects on Competitive Advantages: A Study of Chinese Companies. *Information Resources Management Journal, 30*(3), 41–57. doi:10.4018/IRMJ.2017070103

Bhushan, M., & Yadav, A. (2017). Concept of Cloud Computing in ESB. In R. Bhadoria, N. Chaudhari, G. Tomar, & S. Singh (Eds.), *Exploring Enterprise Service Bus in the Service-Oriented Architecture Paradigm* (pp. 116–127). Hershey, PA: IGI Global. doi:10.4018/978-1-5225-2157-0.ch008

Bhushan, S. (2017). System Dynamics Base-Model of Humanitarian Supply Chain (HSCM) in Disaster Prone Eco-Communities of India: A Discussion on Simulation and Scenario Results. *International Journal of System Dynamics Applications*, *6*(3), 20–37. doi:10.4018/IJSDA.2017070102

Binder, D., & Miller, J. W. (2021). A Generations' Perspective on Employer Branding in Tourism. In V. Costa, A. Moura, & M. Mira (Eds.), *Handbook of Research on Human Capital and People Management in the Tourism Industry* (pp. 152–174). IGI Global. https://doi.org/10.4018/978-1-7998-4318-4.ch008

Birch Freeman, A. A., Mensah, I., & Antwi, K. B. (2022). Smiling vs. Frowning Faces: Community Participation for Sustainable Tourism in Ghanaian Communities. In I. Mensah & E. Afenyo-Agbe (Eds.), *Prospects and Challenges of Community-Based Tourism and Changing Demographics* (pp. 83–106). IGI Global. https://doi.org/10.4018/978-1-7998-7335-8.ch004

Biswas, A., & De, A. K. (2017). On Development of a Fuzzy Stochastic Programming Model with Its Application to Business Management. In S. Trivedi, S. Dey, A. Kumar, & T. Panda (Eds.), *Handbook of Research on Advanced Data Mining Techniques and Applications for Business Intelligence* (pp. 353–378). Hershey, PA: IGI Global. doi:10.4018/978-1-5225-2031-3.ch021

Boragnio, A., & Faracce Macia, C. (2021). "Taking Care of Yourself at Home": Use of E-Commerce About Food and Care During the COVID-19 Pandemic in the City of Buenos Aires. In M. Korstanje (Ed.), *Socio-Economic Effects and Recovery Efforts for the Rental Industry: Post-COVID-19 Strategies* (pp. 45–71). IGI Global. https://doi.org/10.4018/978-1-7998-7287-0.ch003

Borges, V. D. (2021). Happiness: The Basis for Public Policy in Tourism. In A. Perinotto, V. Mayer, & J. Soares (Eds.), *Rebuilding and Restructuring the Tourism Industry: Infusion of Happiness and Quality of Life* (pp. 1–25). IGI Global. https://doi.org/10.4018/978-1-7998-7239-9.ch001

Bücker, J., & Ernste, K. (2018). Use of Brand Heroes in Strategic Reputation Management: The Case of Bacardi, Adidas, and Daimler. In A. Erdemir (Ed.), *Reputation Management Techniques in Public Relations* (pp. 126–150). Hershey, PA: IGI Global. doi:10.4018/978-1-5225-3619-2.ch007

Buluk Eşitti, B. (2021). COVID-19 and Alternative Tourism: New Destinations and New Tourism Products. In M. Demir, A. Dalgıç, & F. Ergen (Eds.), *Handbook of Research on the Impacts and Implications of COVID-19 on the Tourism Industry* (pp. 786–805). IGI Global. https://doi.org/10.4018/978-1-7998-8231-2.ch038

Bureš, V. (2018). Industry 4.0 From the Systems Engineering Perspective: Alternative Holistic Framework Development. In R. Brunet-Thornton & F. Martinez (Eds.), *Analyzing the Impacts of Industry 4.0 in Modern Business Environments* (pp. 199–223). Hershey, PA: IGI Global. doi:10.4018/978-1-5225-3468-6.ch011

Buzady, Z. (2017). Resolving the Magic Cube of Effective Case Teaching: Benchmarking Case Teaching Practices in Emerging Markets – Insights from the Central European University Business School, Hungary. In D. Latusek (Ed.), *Case Studies as a Teaching Tool in Management Education* (pp. 79–103). Hershey, PA: IGI Global. doi:10.4018/978-1-5225-0770-3.ch005

Camillo, A. (2021). *Legal Matters, Risk Management, and Risk Prevention: From Forming a Business to Legal Representation*. IGI Global. doi:10.4018/978-1-7998-4342-9.ch004

Căpusneanu, S., & Topor, D. I. (2018). Business Ethics and Cost Management in SMEs: Theories of Business Ethics and Cost Management Ethos. In I. Oncioiu (Ed.), *Ethics and Decision-Making for Sustainable Business Practices* (pp. 109–127). Hershey, PA: IGI Global. doi:10.4018/978-1-5225-3773-1.ch007

Chan, R. L., Mo, P. L., & Moon, K. K. (2018). Strategic and Tactical Measures in Managing Enterprise Risks: A Study of the Textile and Apparel Industry. In K. Strang, M. Korstanje, & N. Vajjhala (Eds.), *Research, Practices, and Innovations in Global Risk and Contingency Management* (pp. 1–19). Hershey, PA: IGI Global. doi:10.4018/978-1-5225-4754-9.ch001

Charlier, S. D., Burke-Smalley, L. A., & Fisher, S. L. (2018). Undergraduate Programs in the U.S: A Contextual and Content-Based Analysis. In J. Mendy (Ed.), *Teaching Human Resources and Organizational Behavior at the College Level* (pp. 26–57). Hershey, PA: IGI Global. doi:10.4018/978-1-5225-2820-3.ch002

Chumillas, J., Güell, M., & Quer, P. (2022). The Use of ICT in Tourist and Educational Literary Routes: The Role of the Guide. In C. Ramos, S. Quinteiro, & A. Gonçalves (Eds.), *ICT as Innovator Between Tourism and Culture* (pp. 15–29). IGI Global. https://doi.org/10.4018/978-1-7998-8165-0.ch002

Dahlberg, T., Kivijärvi, H., & Saarinen, T. (2017). IT Investment Consistency and Other Factors Influencing the Success of IT Performance. In S. De Haes & W. Van Grembergen (Eds.), *Strategic IT Governance and Alignment in Business Settings* (pp. 176–208). Hershey, PA: IGI Global. doi:10.4018/978-1-5225-0861-8.ch007

Damnjanović, A. M. (2017). Knowledge Management Optimization through IT and E-Business Utilization: A Qualitative Study on Serbian SMEs. In M. Vemić (Ed.), *Optimal Management Strategies in Small and Medium Enterprises* (pp. 249–267). Hershey, PA: IGI Global. doi:10.4018/978-1-5225-1949-2.ch012

Daneshpour, H. (2017). Integrating Sustainable Development into Project Portfolio Management through Application of Open Innovation. In M. Vemić (Ed.), *Optimal Management Strategies in Small and Medium Enterprises* (pp. 370–387). Hershey, PA: IGI Global. doi:10.4018/978-1-5225-1949-2.ch017

Daniel, A. D., & Reis de Castro, V. (2018). Entrepreneurship Education: How to Measure the Impact on Nascent Entrepreneurs. In A. Carrizo Moreira, J. Guilherme Leitão Dantas, & F. Manuel Valente (Eds.), *Nascent Entrepreneurship and Successful New Venture Creation* (pp. 85–110). Hershey, PA: IGI Global. doi:10.4018/978-1-5225-2936-1.ch004

David, R., Swami, B. N., & Tangirala, S. (2018). Ethics Impact on Knowledge Management in Organizational Development: A Case Study. In N. Baporikar (Ed.), *Global Practices in Knowledge Management for Societal and Organizational Development* (pp. 19–45). Hershey, PA: IGI Global. doi:10.4018/978-1-5225-3009-1.ch002

De Uña-Álvarez, E., & Villarino-Pérez, M. (2022). Fostering Ecocultural Resources, Identity, and Tourism in Inland Territories (Galicia, NW Spain). In G. Fernandes (Ed.), *Challenges and New Opportunities for Tourism in Inland Territories: Ecocultural Resources and Sustainable Initiatives* (pp. 1-16). IGI Global. https://doi.org/10.4018/978-1-7998-7339-6.ch001

Delias, P., & Lakiotaki, K. (2018). Discovering Process Horizontal Boundaries to Facilitate Process Comprehension. *International Journal of Operations Research and Information Systems*, *9*(2), 1–31. doi:10.4018/IJORIS.2018040101

Denholm, J., & Lee-Davies, L. (2018). Success Factors for Games in Business and Project Management. In *Enhancing Education and Training Initiatives Through Serious Games* (pp. 34–68). Hershey, PA: IGI Global. doi:10.4018/978-1-5225-3689-5.ch002

Deshpande, M. (2017). Best Practices in Management Institutions for Global Leadership: Policy Aspects. In N. Baporikar (Ed.), *Management Education for Global Leadership* (pp. 1–27). Hershey, PA: IGI Global. doi:10.4018/978-1-5225-1013-0.ch001

Deshpande, M. (2018). Policy Perspectives for SMEs Knowledge Management. In N. Baporikar (Ed.), *Knowledge Integration Strategies for Entrepreneurship and Sustainability* (pp. 23–46). Hershey, PA: IGI Global. doi:10.4018/978-1-5225-5115-7.ch002

Dezdar, S. (2017). ERP Implementation Projects in Asian Countries: A Comparative Study on Iran and China. *International Journal of Information Technology Project Management, 8*(3), 52–68. doi:10.4018/IJITPM.2017070104

Domingos, D., Respício, A., & Martinho, R. (2017). Reliability of IoT-Aware BPMN Healthcare Processes. In C. Reis & M. Maximiano (Eds.), *Internet of Things and Advanced Application in Healthcare* (pp. 214–248). Hershey, PA: IGI Global. doi:10.4018/978-1-5225-1820-4.ch008

Dosumu, O., Hussain, J., & El-Gohary, H. (2017). An Exploratory Study of the Impact of Government Policies on the Development of Small and Medium Enterprises in Developing Countries: The Case of Nigeria. *International Journal of Customer Relationship Marketing and Management, 8*(4), 51–62. doi:10.4018/IJCRMM.2017100104

Durst, S., Bruns, G., & Edvardsson, I. R. (2017). Retaining Knowledge in Smaller Building and Construction Firms. *International Journal of Knowledge and Systems Science, 8*(3), 1–12. doi:10.4018/IJKSS.2017070101

Edvardsson, I. R., & Durst, S. (2017). Outsourcing, Knowledge, and Learning: A Critical Review. *International Journal of Knowledge-Based Organizations, 7*(2), 13–26. doi:10.4018/IJKBO.2017040102

Edwards, J. S. (2018). Integrating Knowledge Management and Business Processes. In M. Khosrow-Pour, D.B.A. (Ed.), Encyclopedia of Information Science and Technology, Fourth Edition (pp. 5046-5055). Hershey, PA: IGI Global. doi:10.4018/978-1-5225-2255-3.ch437

Eichelberger, S., & Peters, M. (2021). Family Firm Management in Turbulent Times: Opportunities for Responsible Tourism. In A. Zehrer, G. Glowka, K. Schwaiger, & V. Ranacher-Lackner (Eds.), *Resiliency Models and Addressing Future Risks for Family Firms in the Tourism Industry* (pp. 103–124). IGI Global. https://doi.org/10.4018/978-1-7998-7352-5.ch005

Eide, D., Hjalager, A., & Hansen, M. (2022). Innovative Certifications in Adventure Tourism: Attributes and Diffusion. In R. Augusto Costa, F. Brandão, Z. Breda, & C. Costa (Eds.), *Planning and Managing the Experience Economy in Tourism* (pp. 161-175). IGI Global. https://doi.org/10.4018/978-1-7998-8775-1.ch009

Ejiogu, A. O. (2018). Economics of Farm Management. In *Agricultural Finance and Opportunities for Investment and Expansion* (pp. 56–72). Hershey, PA: IGI Global. doi:10.4018/978-1-5225-3059-6.ch003

Ekanem, I., & Abiade, G. E. (2018). Factors Influencing the Use of E-Commerce by Small Enterprises in Nigeria. *International Journal of ICT Research in Africa and the Middle East*, *7*(1), 37–53. doi:10.4018/IJICTRAME.2018010103

Ekanem, I., & Alrossais, L. A. (2017). Succession Challenges Facing Family Businesses in Saudi Arabia. In P. Zgheib (Ed.), *Entrepreneurship and Business Innovation in the Middle East* (pp. 122–146). Hershey, PA: IGI Global. doi:10.4018/978-1-5225-2066-5.ch007

El Faquih, L., & Fredj, M. (2017). Ontology-Based Framework for Quality in Configurable Process Models. *Journal of Electronic Commerce in Organizations*, *15*(2), 48–60. doi:10.4018/JECO.2017040104

Faisal, M. N., & Talib, F. (2017). Building Ambidextrous Supply Chains in SMEs: How to Tackle the Barriers? *International Journal of Information Systems and Supply Chain Management*, *10*(4), 80–100. doi:10.4018/IJISSCM.2017100105

Fernandes, T. M., Gomes, J., & Romão, M. (2017). Investments in E-Government: A Benefit Management Case Study. *International Journal of Electronic Government Research*, *13*(3), 1–17. doi:10.4018/IJEGR.2017070101

Figueira, L. M., Honrado, G. R., & Dionísio, M. S. (2021). Human Capital Management in the Tourism Industry in Portugal. In V. Costa, A. Moura, & M. Mira (Eds.), *Handbook of Research on Human Capital and People Management in the Tourism Industry* (pp. 1–19). IGI Global. doi:10.4018/978-1-7998-4318-4.ch001

Gao, S. S., Oreal, S., & Zhang, J. (2018). Contemporary Financial Risk Management Perceptions and Practices of Small-Sized Chinese Businesses. In I. Management Association (Ed.), Global Business Expansion: Concepts, Methodologies, Tools, and Applications (pp. 917-931). Hershey, PA: IGI Global. doi:10.4018/978-1-5225-5481-3.ch041

Garg, R., & Berning, S. C. (2017). Indigenous Chinese Management Philosophies: Key Concepts and Relevance for Modern Chinese Firms. In B. Christiansen & G. Koc (Eds.), *Transcontinental Strategies for Industrial Development and Economic Growth* (pp. 43–57). Hershey, PA: IGI Global. doi:10.4018/978-1-5225-2160-0.ch003

Gencer, Y. G. (2017). Supply Chain Management in Retailing Business. In U. Akkucuk (Ed.), *Ethics and Sustainability in Global Supply Chain Management* (pp. 197–210). Hershey, PA: IGI Global. doi:10.4018/978-1-5225-2036-8.ch011

Gera, R., Arora, S., & Malik, S. (2021). Emotional Labor in the Tourism Industry: Strategies, Antecedents, and Outcomes. In V. Costa, A. Moura, & M. Mira (Eds.), *Handbook of Research on Human Capital and People Management in the Tourism Industry* (pp. 73–91). IGI Global. https://doi.org/10.4018/978-1-7998-4318-4.ch004

Giacosa, E. (2018). The Increasing of the Regional Development Thanks to the Luxury Business Innovation. In L. Carvalho (Ed.), *Handbook of Research on Entrepreneurial Ecosystems and Social Dynamics in a Globalized World* (pp. 260–273). Hershey, PA: IGI Global. doi:10.4018/978-1-5225-3525-6.ch011

Glowka, G., Tusch, M., & Zehrer, A. (2021). The Risk Perception of Family Business Owner-Manager in the Tourism Industry: A Qualitative Comparison of the Intra-Firm Senior and Junior Generation. In A. Zehrer, G. Glowka, K. Schwaiger, & V. Ranacher-Lackner (Eds.), *Resiliency Models and Addressing Future Risks for Family Firms in the Tourism Industry* (pp. 126–153). IGI Global. https://doi.org/10.4018/978-1-7998-7352-5.ch006

Glykas, M., & George, J. (2017). Quality and Process Management Systems in the UAE Maritime Industry. *International Journal of Productivity Management and Assessment Technologies*, *5*(1), 20–39. doi:10.4018/IJPMAT.2017010102

Glykas, M., Valiris, G., Kokkinaki, A., & Koutsoukou, Z. (2018). Banking Business Process Management Implementation. *International Journal of Productivity Management and Assessment Technologies*, *6*(1), 50–69. doi:10.4018/IJPMAT.2018010104

Gomes, J., & Romão, M. (2017). The Balanced Scorecard: Keeping Updated and Aligned with Today's Business Trends. *International Journal of Productivity Management and Assessment Technologies*, *5*(2), 1–15. doi:10.4018/IJPMAT.2017070101

Gomes, J., & Romão, M. (2017). Aligning Information Systems and Technology with Benefit Management and Balanced Scorecard. In S. De Haes & W. Van Grembergen (Eds.), *Strategic IT Governance and Alignment in Business Settings* (pp. 112–131). Hershey, PA: IGI Global. doi:10.4018/978-1-5225-0861-8.ch005

Goyal, A. (2021). Communicating and Building Destination Brands With New Media. In M. Dinis, L. Bonixe, S. Lamy, & Z. Breda (Eds.), *Impact of New Media in Tourism* (pp. 1-20). IGI Global. https://doi.org/10.4018/978-1-7998-7095-1.ch001

Grefen, P., & Turetken, O. (2017). Advanced Business Process Management in Networked E-Business Scenarios. *International Journal of E-Business Research, 13*(4), 70–104. doi:10.4018/IJEBR.2017100105

Guasca, M., Van Broeck, A. M., & Vanneste, D. (2021). Tourism and the Social Reintegration of Colombian Ex-Combatants. In J. da Silva, Z. Breda, & F. Carbone (Eds.), *Role and Impact of Tourism in Peacebuilding and Conflict Transformation* (pp. 66-86). IGI Global. https://doi.org/10.4018/978-1-7998-5053-3.ch005

Haider, A., & Saetang, S. (2017). Strategic IT Alignment in Service Sector. In S. Rozenes & Y. Cohen (Eds.), *Handbook of Research on Strategic Alliances and Value Co-Creation in the Service Industry* (pp. 231–258). Hershey, PA: IGI Global. doi:10.4018/978-1-5225-2084-9.ch012

Hajilari, A. B., Ghadaksaz, M., & Fasghandis, G. S. (2017). Assessing Organizational Readiness for Implementing ERP System Using Fuzzy Expert System Approach. *International Journal of Enterprise Information Systems, 13*(1), 67–85. doi:10.4018/IJEIS.2017010105

Haldorai, A., Ramu, A., & Murugan, S. (2018). Social Aware Cognitive Radio Networks: Effectiveness of Social Networks as a Strategic Tool for Organizational Business Management. In H. Bansal, G. Shrivastava, G. Nguyen, & L. Stanciu (Eds.), *Social Network Analytics for Contemporary Business Organizations* (pp. 188–202). Hershey, PA: IGI Global. doi:10.4018/978-1-5225-5097-6.ch010

Hall, O. P. Jr. (2017). Social Media Driven Management Education. *International Journal of Knowledge-Based Organizations, 7*(2), 43–59. doi:10.4018/IJKBO.2017040104

Hanifah, H., Halim, H. A., Ahmad, N. H., & Vafaei-Zadeh, A. (2017). Innovation Culture as a Mediator Between Specific Human Capital and Innovation Performance Among Bumiputera SMEs in Malaysia. In N. Ahmad, T. Ramayah, H. Halim, & S. Rahman (Eds.), *Handbook of Research on Small and Medium Enterprises in Developing Countries* (pp. 261–279). Hershey, PA: IGI Global. doi:10.4018/978-1-5225-2165-5.ch012

Hartlieb, S., & Silvius, G. (2017). Handling Uncertainty in Project Management and Business Development: Similarities and Differences. In Y. Raydugin (Ed.), *Handbook of Research on Leveraging Risk and Uncertainties for Effective Project Management* (pp. 337–362). Hershey, PA: IGI Global. doi:10.4018/978-1-5225-1790-0.ch016

Hass, K. B. (2017). Living on the Edge: Managing Project Complexity. In Y. Raydugin (Ed.), *Handbook of Research on Leveraging Risk and Uncertainties for Effective Project Management* (pp. 177–201). Hershey, PA: IGI Global. doi:10.4018/978-1-5225-1790-0.ch009

Hawking, P., & Carmine Sellitto, C. (2017). Developing an Effective Strategy for Organizational Business Intelligence. In M. Tavana (Ed.), *Enterprise Information Systems and the Digitalization of Business Functions* (pp. 222–237). Hershey, PA: IGI Global. doi:10.4018/978-1-5225-2382-6.ch010

Hawking, P., & Sellitto, C. (2017). A Fast-Moving Consumer Goods Company and Business Intelligence Strategy Development. *International Journal of Enterprise Information Systems, 13*(2), 22–33. doi:10.4018/IJEIS.2017040102

Hawking, P., & Sellitto, C. (2017). Business Intelligence Strategy: Two Case Studies. *International Journal of Business Intelligence Research, 8*(2), 17–30. doi:10.4018/IJBIR.2017070102

Hee, W. J., Jalleh, G., Lai, H., & Lin, C. (2017). E-Commerce and IT Projects: Evaluation and Management Issues in Australian and Taiwanese Hospitals. *International Journal of Public Health Management and Ethics, 2*(1), 69–90. doi:10.4018/IJPHME.2017010104

Hernandez, A. A. (2018). Exploring the Factors to Green IT Adoption of SMEs in the Philippines. *Journal of Cases on Information Technology, 20*(2), 49–66. doi:10.4018/JCIT.2018040104

Hollman, A., Bickford, S., & Hollman, T. (2017). Cyber InSecurity: A Post-Mortem Attempt to Assess Cyber Problems from IT and Business Management Perspectives. *Journal of Cases on Information Technology, 19*(3), 42–70. doi:10.4018/JCIT.2017070104

Ibrahim, F., & Zainin, N. M. (2021). Exploring the Technological Impacts: The Case of Museums in Brunei Darussalam. *International Journal of Tourism and Hospitality Management in the Digital Age, 5*(1), 1–18. https://doi.org/10.4018/IJTHMDA.2021010101

Igbinakhase, I. (2017). Responsible and Sustainable Management Practices in Developing and Developed Business Environments. In Z. Fields (Ed.), *Collective Creativity for Responsible and Sustainable Business Practice* (pp. 180–207). Hershey, PA: IGI Global. doi:10.4018/978-1-5225-1823-5.ch010

Iwata, J. J., & Hoskins, R. G. (2017). Managing Indigenous Knowledge in Tanzania: A Business Perspective. In P. Jain & N. Mnjama (Eds.), *Managing Knowledge Resources and Records in Modern Organizations* (pp. 198–214). Hershey, PA: IGI Global. doi:10.4018/978-1-5225-1965-2.ch012

Jain, P. (2017). Ethical and Legal Issues in Knowledge Management Life-Cycle in Business. In P. Jain & N. Mnjama (Eds.), *Managing Knowledge Resources and Records in Modern Organizations* (pp. 82–101). Hershey, PA: IGI Global. doi:10.4018/978-1-5225-1965-2.ch006

James, S., & Hauli, E. (2017). Holistic Management Education at Tanzanian Rural Development Planning Institute. In N. Baporikar (Ed.), *Management Education for Global Leadership* (pp. 112–136). Hershey, PA: IGI Global. doi:10.4018/978-1-5225-1013-0.ch006

Janošková, M., Csikósová, A., & Čulková, K. (2018). Measurement of Company Performance as Part of Its Strategic Management. In R. Leon (Ed.), *Managerial Strategies for Business Sustainability During Turbulent Times* (pp. 309–335). Hershey, PA: IGI Global. doi:10.4018/978-1-5225-2716-9.ch017

Jean-Vasile, A., & Alecu, A. (2017). Theoretical and Practical Approaches in Understanding the Influences of Cost-Productivity-Profit Trinomial in Contemporary Enterprises. In A. Jean Vasile & D. Nicolò (Eds.), *Sustainable Entrepreneurship and Investments in the Green Economy* (pp. 28–62). Hershey, PA: IGI Global. doi:10.4018/978-1-5225-2075-7.ch002

Joia, L. A., & Correia, J. C. (2018). CIO Competencies From the IT Professional Perspective: Insights From Brazil. *Journal of Global Information Management, 26*(2), 74–103. doi:10.4018/JGIM.2018040104

Juma, A., & Mzera, N. (2017). Knowledge Management and Records Management and Competitive Advantage in Business. In P. Jain & N. Mnjama (Eds.), *Managing Knowledge Resources and Records in Modern Organizations* (pp. 15–28). Hershey, PA: IGI Global. doi:10.4018/978-1-5225-1965-2.ch002

K., I., & A, V. (2018). Monitoring and Auditing in the Cloud. In K. Munir (Ed.), *Cloud Computing Technologies for Green Enterprises* (pp. 318-350). Hershey, PA: IGI Global. https://doi.org/ doi:10.4018/978-1-5225-3038-1.ch013

Kabra, G., Ghosh, V., & Ramesh, A. (2018). Enterprise Integrated Business Process Management and Business Intelligence Framework for Business Process Sustainability. In A. Paul, D. Bhattacharyya, & S. Anand (Eds.), *Green Initiatives for Business Sustainability and Value Creation* (pp. 228–238). Hershey, PA: IGI Global. doi:10.4018/978-1-5225-2662-9.ch010

Kaoud, M. (2017). Investigation of Customer Knowledge Management: A Case Study Research. *International Journal of Service Science, Management, Engineering, and Technology*, *8*(2), 12–22. doi:10.4018/IJSSMET.2017040102

Katuu, S. (2018). A Comparative Assessment of Enterprise Content Management Maturity Models. In N. Gwangwava & M. Mutingi (Eds.), *E-Manufacturing and E-Service Strategies in Contemporary Organizations* (pp. 93–118). Hershey, PA: IGI Global. doi:10.4018/978-1-5225-3628-4.ch005

Khan, M. Y., & Abir, T. (2022). The Role of Social Media Marketing in the Tourism and Hospitality Industry: A Conceptual Study on Bangladesh. In C. Ramos, S. Quinteiro, & A. Gonçalves (Eds.), *ICT as Innovator Between Tourism and Culture* (pp. 213–229). IGI Global. https://doi.org/10.4018/978-1-7998-8165-0.ch013

Kinnunen, S., Ylä-Kujala, A., Marttonen-Arola, S., Kärri, T., & Baglee, D. (2018). Internet of Things in Asset Management: Insights from Industrial Professionals and Academia. *International Journal of Service Science, Management, Engineering, and Technology*, *9*(2), 104–119. doi:10.4018/IJSSMET.2018040105

Klein, A. Z., Sabino de Freitas, A., Machado, L., Freitas, J. C. Jr, Graziola, P. G. Jr, & Schlemmer, E. (2017). Virtual Worlds Applications for Management Education. In L. Tomei (Ed.), *Exploring the New Era of Technology-Infused Education* (pp. 279–299). Hershey, PA: IGI Global. doi:10.4018/978-1-5225-1709-2.ch017

Kővári, E., Saleh, M., & Steinbachné Hajmásy, G. (2022). The Impact of Corporate Digital Responsibility (CDR) on Internal Stakeholders' Satisfaction in Hungarian Upscale Hotels. In M. Valeri (Ed.), *New Governance and Management in Touristic Destinations* (pp. 35–51). IGI Global. https://doi.org/10.4018/978-1-6684-3889-3.ch003

Kożuch, B., & Jabłoński, A. (2017). Adopting the Concept of Business Models in Public Management. In M. Lewandowski & B. Kożuch (Eds.), *Public Sector Entrepreneurship and the Integration of Innovative Business Models* (pp. 10–46). Hershey, PA: IGI Global. doi:10.4018/978-1-5225-2215-7.ch002

Kumar, J., Adhikary, A., & Jha, A. (2017). Small Active Investors' Perceptions and Preferences Towards Tax Saving Mutual Fund Schemes in Eastern India: An Empirical Note. *International Journal of Asian Business and Information Management*, *8*(2), 35–45. doi:10.4018/IJABIM.2017040103

Latusi, S., & Fissore, M. (2021). Pilgrimage Routes to Happiness: Comparing the Camino de Santiago and Via Francigena. In A. Perinotto, V. Mayer, & J. Soares (Eds.), *Rebuilding and Restructuring the Tourism Industry: Infusion of Happiness and Quality of Life* (pp. 157–182). IGI Global. https://doi.org/10.4018/978-1-7998-7239-9.ch008

Lavassani, K. M., & Movahedi, B. (2017). Applications Driven Information Systems: Beyond Networks toward Business Ecosystems. *International Journal of Innovation in the Digital Economy*, *8*(1), 61–75. doi:10.4018/IJIDE.2017010104

Lazzareschi, V. H., & Brito, M. S. (2017). Strategic Information Management: Proposal of Business Project Model. In G. Jamil, A. Soares, & C. Pessoa (Eds.), *Handbook of Research on Information Management for Effective Logistics and Supply Chains* (pp. 59–88). Hershey, PA: IGI Global. doi:10.4018/978-1-5225-0973-8.ch004

Lechuga Sancho, M. P., & Martín Navarro, A. (2022). Evolution of the Literature on Social Responsibility in the Tourism Sector: A Systematic Literature Review. In G. Fernandes (Ed.), *Challenges and New Opportunities for Tourism in Inland Territories: Ecocultural Resources and Sustainable Initiatives* (pp. 169–186). IGI Global. https://doi.org/10.4018/978-1-7998-7339-6.ch010

Lederer, M., Kurz, M., & Lazarov, P. (2017). Usage and Suitability of Methods for Strategic Business Process Initiatives: A Multi Case Study Research. *International Journal of Productivity Management and Assessment Technologies*, *5*(1), 40–51. doi:10.4018/IJPMAT.2017010103

Lee, I. (2017). A Social Enterprise Business Model and a Case Study of Pacific Community Ventures (PCV). In V. Potocan, M. Ünğan, & Z. Nedelko (Eds.), *Handbook of Research on Managerial Solutions in Non-Profit Organizations* (pp. 182–204). Hershey, PA: IGI Global. doi:10.4018/978-1-5225-0731-4.ch009

Leon, L. A., Seal, K. C., Przasnyski, Z. H., & Wiedenman, I. (2017). Skills and Competencies Required for Jobs in Business Analytics: A Content Analysis of Job Advertisements Using Text Mining. *International Journal of Business Intelligence Research*, *8*(1), 1–25. doi:10.4018/IJBIR.2017010101

Levy, C. L., & Elias, N. I. (2017). SOHO Users' Perceptions of Reliability and Continuity of Cloud-Based Services. In M. Moore (Ed.), *Cybersecurity Breaches and Issues Surrounding Online Threat Protection* (pp. 248–287). Hershey, PA: IGI Global. doi:10.4018/978-1-5225-1941-6.ch011

Levy, M. (2018). Change Management Serving Knowledge Management and Organizational Development: Reflections and Review. In N. Baporikar (Ed.), *Global Practices in Knowledge Management for Societal and Organizational Development* (pp. 256–270). Hershey, PA: IGI Global. doi:10.4018/978-1-5225-3009-1.ch012

Lewandowski, M. (2017). Public Organizations and Business Model Innovation: The Role of Public Service Design. In M. Lewandowski & B. Kożuch (Eds.), *Public Sector Entrepreneurship and the Integration of Innovative Business Models* (pp. 47–72). Hershey, PA: IGI Global. doi:10.4018/978-1-5225-2215-7.ch003

Lhannaoui, H., Kabbaj, M. I., & Bakkoury, Z. (2017). A Survey of Risk-Aware Business Process Modelling. *International Journal of Risk and Contingency Management, 6*(3), 14–26. doi:10.4018/IJRCM.2017070102

Li, J., Sun, W., Jiang, W., Yang, H., & Zhang, L. (2017). How the Nature of Exogenous Shocks and Crises Impact Company Performance?: The Effects of Industry Characteristics. *International Journal of Risk and Contingency Management, 6*(4), 40–55. doi:10.4018/IJRCM.2017100103

Lopez-Fernandez, M., Perez-Perez, M., Serrano-Bedia, A., & Cobo-Gonzalez, A. (2021). Small and Medium Tourism Enterprise Survival in Times of Crisis: "El Capricho de Gaudí. In D. Toubes & N. Araújo-Vila (Eds.), *Risk, Crisis, and Disaster Management in Small and Medium-Sized Tourism Enterprises* (pp. 103–129). IGI Global. doi:10.4018/978-1-7998-6996-2.ch005

Mahajan, A., Maidullah, S., & Hossain, M. R. (2022). Experience Toward Smart Tour Guide Apps in Travelling: An Analysis of Users' Reviews on Audio Odigos and Trip My Way. In R. Augusto Costa, F. Brandão, Z. Breda, & C. Costa (Eds.), *Planning and Managing the Experience Economy in Tourism* (pp. 255-273). IGI Global. https://doi.org/10.4018/978-1-7998-8775-1.ch014

Malega, P. (2017). Small and Medium Enterprises in the Slovak Republic: Status and Competitiveness of SMEs in the Global Markets and Possibilities of Optimization. In M. Vemić (Ed.), *Optimal Management Strategies in Small and Medium Enterprises* (pp. 102–124). Hershey, PA: IGI Global. doi:10.4018/978-1-5225-1949-2.ch006

Malewska, K. M. (2017). Intuition in Decision-Making on the Example of a Non-Profit Organization. In V. Potocan, M. Üngan, & Z. Nedelko (Eds.), *Handbook of Research on Managerial Solutions in Non-Profit Organizations* (pp. 378–399). Hershey, PA: IGI Global. doi:10.4018/978-1-5225-0731-4.ch018

Maroofi, F. (2017). Entrepreneurial Orientation and Organizational Learning Ability Analysis for Innovation and Firm Performance. In N. Baporikar (Ed.), *Innovation and Shifting Perspectives in Management Education* (pp. 144–165). Hershey, PA: IGI Global. doi:10.4018/978-1-5225-1019-2.ch007

Marques, M., Moleiro, D., Brito, T. M., & Marques, T. (2021). Customer Relationship Management as an Important Relationship Marketing Tool: The Case of the Hospitality Industry in Estoril Coast. In M. Dinis, L. Bonixe, S. Lamy, & Z. Breda (Eds.), Impact of New Media in Tourism (pp. 39-56). IGI Global. https://doi.org/doi:10.4018/978-1-7998-7095-1.ch003

Martins, P. V., & Zacarias, M. (2017). A Web-based Tool for Business Process Improvement. *International Journal of Web Portals, 9*(2), 68–84. doi:10.4018/IJWP.2017070104

Matthies, B., & Coners, A. (2017). Exploring the Conceptual Nature of e-Business Projects. *Journal of Electronic Commerce in Organizations, 15*(3), 33–63. doi:10.4018/JECO.2017070103

Mayer, V. F., Fraga, C. C., & Silva, L. C. (2021). Contributions of Neurosciences to Studies of Well-Being in Tourism. In A. Perinotto, V. Mayer, & J. Soares (Eds.), *Rebuilding and Restructuring the Tourism Industry: Infusion of Happiness and Quality of Life* (pp. 108–128). IGI Global. https://doi.org/10.4018/978-1-7998-7239-9.ch006

McKee, J. (2018). Architecture as a Tool to Solve Business Planning Problems. In M. Khosrow-Pour, D.B.A. (Ed.), Encyclopedia of Information Science and Technology, Fourth Edition (pp. 573-586). Hershey, PA: IGI Global. doi:10.4018/978-1-5225-2255-3.ch050

McMurray, A. J., Cross, J., & Caponecchia, C. (2018). The Risk Management Profession in Australia: Business Continuity Plan Practices. In N. Bajgoric (Ed.), *Always-On Enterprise Information Systems for Modern Organizations* (pp. 112–129). Hershey, PA: IGI Global. doi:10.4018/978-1-5225-3704-5.ch006

Meddah, I. H., & Belkadi, K. (2018). Mining Patterns Using Business Process Management. In R. Hamou (Ed.), *Handbook of Research on Biomimicry in Information Retrieval and Knowledge Management* (pp. 78–89). Hershey, PA: IGI Global. doi:10.4018/978-1-5225-3004-6.ch005

Melian, A. G., & Camprubí, R. (2021). The Accessibility of Museum Websites: The Case of Barcelona. In C. Eusébio, L. Teixeira, & M. Carneiro (Eds.), *ICT Tools and Applications for Accessible Tourism* (pp. 234–255). IGI Global. https://doi.org/10.4018/978-1-7998-6428-8.ch011

Mendes, L. (2017). TQM and Knowledge Management: An Integrated Approach Towards Tacit Knowledge Management. In D. Jaziri-Bouagina & G. Jamil (Eds.), *Handbook of Research on Tacit Knowledge Management for Organizational Success* (pp. 236–263). Hershey, PA: IGI Global. doi:10.4018/978-1-5225-2394-9.ch009

Menezes, V. D., & Cavagnaro, E. (2021). Communicating Sustainable Initiatives in the Hotel Industry: The Case of the Hotel Jakarta Amsterdam. In F. Brandão, Z. Breda, R. Costa, & C. Costa (Eds.), *Handbook of Research on the Role of Tourism in Achieving Sustainable Development Goals* (pp. 224-234). IGI Global. https://doi.org/10.4018/978-1-7998-5691-7.ch013

Menezes, V. D., & Cavagnaro, E. (2021). Communicating Sustainable Initiatives in the Hotel Industry: The Case of the Hotel Jakarta Amsterdam. In F. Brandão, Z. Breda, R. Costa, & C. Costa (Eds.), *Handbook of Research on the Role of Tourism in Achieving Sustainable Development Goals* (pp. 224-234). IGI Global. https://doi.org/10.4018/978-1-7998-5691-7.ch013

Mitas, O., Bastiaansen, M., & Boode, W. (2022). If You're Happy, I'm Happy: Emotion Contagion at a Tourist Information Center. In R. Augusto Costa, F. Brandão, Z. Breda, & C. Costa (Eds.), *Planning and Managing the Experience Economy in Tourism* (pp. 122-140). IGI Global. https://doi.org/10.4018/978-1-7998-8775-1.ch007

Mnjama, N. M. (2017). Preservation of Recorded Information in Public and Private Sector Organizations. In P. Jain & N. Mnjama (Eds.), *Managing Knowledge Resources and Records in Modern Organizations* (pp. 149–167). Hershey, PA: IGI Global. doi:10.4018/978-1-5225-1965-2.ch009

Mokoqama, M., & Fields, Z. (2017). Principles of Responsible Management Education (PRME): Call for Responsible Management Education. In Z. Fields (Ed.), *Collective Creativity for Responsible and Sustainable Business Practice* (pp. 229–241). Hershey, PA: IGI Global. doi:10.4018/978-1-5225-1823-5.ch012

Monteiro, A., Lopes, S., & Carbone, F. (2021). Academic Mobility: Bridging Tourism and Peace Education. In J. da Silva, Z. Breda, & F. Carbone (Eds.), *Role and Impact of Tourism in Peacebuilding and Conflict Transformation* (pp. 275-301). IGI Global. https://doi.org/10.4018/978-1-7998-5053-3.ch016

Muniapan, B. (2017). Philosophy and Management: The Relevance of Vedanta in Management. In P. Ordóñez de Pablos (Ed.), *Managerial Strategies and Solutions for Business Success in Asia* (pp. 124–139). Hershey, PA: IGI Global. doi:10.4018/978-1-5225-1886-0.ch007

Murad, S. E., & Dowaji, S. (2017). Using Value-Based Approach for Managing Cloud-Based Services. In A. Turuk, B. Sahoo, & S. Addya (Eds.), *Resource Management and Efficiency in Cloud Computing Environments* (pp. 33–60). Hershey, PA: IGI Global. doi:10.4018/978-1-5225-1721-4.ch002

Mutahar, A. M., Daud, N. M., Thurasamy, R., Isaac, O., & Abdulsalam, R. (2018). The Mediating of Perceived Usefulness and Perceived Ease of Use: The Case of Mobile Banking in Yemen. *International Journal of Technology Diffusion*, *9*(2), 21–40. doi:10.4018/IJTD.2018040102

Naidoo, V. (2017). E-Learning and Management Education at African Universities. In N. Baporikar (Ed.), *Management Education for Global Leadership* (pp. 181–201). Hershey, PA: IGI Global. doi:10.4018/978-1-5225-1013-0.ch009

Naidoo, V., & Igbinakhase, I. (2018). Opportunities and Challenges of Knowledge Retention in SMEs. In N. Baporikar (Ed.), *Knowledge Integration Strategies for Entrepreneurship and Sustainability* (pp. 70–94). Hershey, PA: IGI Global. doi:10.4018/978-1-5225-5115-7.ch004

Naumov, N., & Costandachi, G. (2021). Creativity and Entrepreneurship: Gastronomic Tourism in Mexico. In J. Soares (Ed.), *Innovation and Entrepreneurial Opportunities in Community Tourism* (pp. 90–108). IGI Global. https://doi.org/10.4018/978-1-7998-4855-4.ch006

Nayak, S., & Prabhu, N. (2017). Paradigm Shift in Management Education: Need for a Cross Functional Perspective. In N. Baporikar (Ed.), *Management Education for Global Leadership* (pp. 241–255). Hershey, PA: IGI Global. doi:10.4018/978-1-5225-1013-0.ch012

Nedelko, Z., & Potocan, V. (2017). Management Solutions in Non-Profit Organizations: Case of Slovenia. In V. Potocan, M. Üngan, & Z. Nedelko (Eds.), *Handbook of Research on Managerial Solutions in Non-Profit Organizations* (pp. 1–22). Hershey, PA: IGI Global. doi:10.4018/978-1-5225-0731-4.ch001

Nedelko, Z., & Potocan, V. (2017). Priority of Management Tools Utilization among Managers: International Comparison. In V. Wang (Ed.), *Encyclopedia of Strategic Leadership and Management* (pp. 1083–1094). Hershey, PA: IGI Global. doi:10.4018/978-1-5225-1049-9.ch075

Nedelko, Z., Raudeliūnienė, J., & Črešnar, R. (2018). Knowledge Dynamics in Supply Chain Management. In N. Baporikar (Ed.), *Knowledge Integration Strategies for Entrepreneurship and Sustainability* (pp. 150–166). Hershey, PA: IGI Global. doi:10.4018/978-1-5225-5115-7.ch008

Nguyen, H. T., & Hipsher, S. A. (2018). Innovation and Creativity Used by Private Sector Firms in a Resources-Constrained Environment. In S. Hipsher (Ed.), *Examining the Private Sector's Role in Wealth Creation and Poverty Reduction* (pp. 219–238). Hershey, PA: IGI Global. doi:10.4018/978-1-5225-3117-3.ch010

Obicci, P. A. (2017). Risk Sharing in a Partnership. In *Risk Management Strategies in Public-Private Partnerships* (pp. 115–152). Hershey, PA: IGI Global. doi:10.4018/978-1-5225-2503-5.ch004

Obidallah, W. J., & Raahemi, B. (2017). Managing Changes in Service Oriented Virtual Organizations: A Structural and Procedural Framework to Facilitate the Process of Change. *Journal of Electronic Commerce in Organizations*, *15*(1), 59–83. doi:10.4018/JECO.2017010104

Ojo, O. (2017). Impact of Innovation on the Entrepreneurial Success in Selected Business Enterprises in South-West Nigeria. *International Journal of Innovation in the Digital Economy*, *8*(2), 29–38. doi:10.4018/IJIDE.2017040103

Okdinawati, L., Simatupang, T. M., & Sunitiyoso, Y. (2017). Multi-Agent Reinforcement Learning for Value Co-Creation of Collaborative Transportation Management (CTM). *International Journal of Information Systems and Supply Chain Management*, *10*(3), 84–95. doi:10.4018/IJISSCM.2017070105

Olivera, V. A., & Carrillo, I. M. (2021). Organizational Culture: A Key Element for the Development of Mexican Micro and Small Tourist Companies. In J. Soares (Ed.), *Innovation and Entrepreneurial Opportunities in Community Tourism* (pp. 227–242). IGI Global. doi:10.4018/978-1-7998-4855-4.ch013

Ossorio, M. (2022). Corporate Museum Experiences in Enogastronomic Tourism. In R. Augusto Costa, F. Brandão, Z. Breda, & C. Costa (Eds.), Planning and Managing the Experience Economy in Tourism (pp. 107-121). IGI Global. https://doi.org/doi:10.4018/978-1-7998-8775-1.ch006

Ossorio, M. (2022). Enogastronomic Tourism in Times of Pandemic. In G. Fernandes (Ed.), *Challenges and New Opportunities for Tourism in Inland Territories: Ecocultural Resources and Sustainable Initiatives* (pp. 241–255). IGI Global. https://doi.org/10.4018/978-1-7998-7339-6.ch014

Özekici, Y. K. (2022). ICT as an Acculturative Agent and Its Role in the Tourism Context: Introduction, Acculturation Theory, Progress of the Acculturation Theory in Extant Literature. In C. Ramos, S. Quinteiro, & A. Gonçalves (Eds.), *ICT as Innovator Between Tourism and Culture* (pp. 42–66). IGI Global. https://doi.org/10.4018/978-1-7998-8165-0.ch004

Pal, K. (2018). Building High Quality Big Data-Based Applications in Supply Chains. In A. Kumar & S. Saurav (Eds.), *Supply Chain Management Strategies and Risk Assessment in Retail Environments* (pp. 1–24). Hershey, PA: IGI Global. doi:10.4018/978-1-5225-3056-5.ch001

Palos-Sanchez, P. R., & Correia, M. B. (2018). Perspectives of the Adoption of Cloud Computing in the Tourism Sector. In J. Rodrigues, C. Ramos, P. Cardoso, & C. Henriques (Eds.), *Handbook of Research on Technological Developments for Cultural Heritage and eTourism Applications* (pp. 377–400). Hershey, PA: IGI Global. doi:10.4018/978-1-5225-2927-9.ch018

Papadopoulou, G. (2021). Promoting Gender Equality and Women Empowerment in the Tourism Sector. In F. Brandão, Z. Breda, R. Costa, & C. Costa (Eds.), Handbook of Research on the Role of Tourism in Achieving Sustainable Development Goals (pp. 152-174). IGI Global. https://doi.org/ doi:10.4018/978-1-7998-5691-7.ch009

Papp-Váry, Á. F., & Tóth, T. Z. (2022). Analysis of Budapest as a Film Tourism Destination. In R. Baleiro & R. Pereira (Eds.), *Global Perspectives on Literary Tourism and Film-Induced Tourism* (pp. 257-279). IGI Global. https://doi.org/10.4018/978-1-7998-8262-6.ch014

Patiño, B. E. (2017). New Generation Management by Convergence and Individual Identity: A Systemic and Human-Oriented Approach. In N. Baporikar (Ed.), *Innovation and Shifting Perspectives in Management Education* (pp. 119–143). Hershey, PA: IGI Global. doi:10.4018/978-1-5225-1019-2.ch006

Patro, C. S. (2021). Digital Tourism: Influence of E-Marketing Technology. In M. Dinis, L. Bonixe, S. Lamy, & Z. Breda (Eds.), *Impact of New Media in Tourism* (pp. 234-254). IGI Global. https://doi.org/10.4018/978-1-7998-7095-1.ch014

Pawliczek, A., & Rössler, M. (2017). Knowledge of Management Tools and Systems in SMEs: Knowledge Transfer in Management. In A. Bencsik (Ed.), *Knowledge Management Initiatives and Strategies in Small and Medium Enterprises* (pp. 180–203). Hershey, PA: IGI Global. doi:10.4018/978-1-5225-1642-2.ch009

Pejic-Bach, M., Omazic, M. A., Aleksic, A., & Zoroja, J. (2018). Knowledge-Based Decision Making: A Multi-Case Analysis. In R. Leon (Ed.), *Managerial Strategies for Business Sustainability During Turbulent Times* (pp. 160–184). Hershey, PA: IGI Global. doi:10.4018/978-1-5225-2716-9.ch009

Perano, M., Hysa, X., & Calabrese, M. (2018). Strategic Planning, Cultural Context, and Business Continuity Management: Business Cases in the City of Shkoder. In A. Presenza & L. Sheehan (Eds.), *Geopolitics and Strategic Management in the Global Economy* (pp. 57–77). Hershey, PA: IGI Global. doi:10.4018/978-1-5225-2673-5.ch004

Pereira, R., Mira da Silva, M., & Lapão, L. V. (2017). IT Governance Maturity Patterns in Portuguese Healthcare. In S. De Haes & W. Van Grembergen (Eds.), *Strategic IT Governance and Alignment in Business Settings* (pp. 24–52). Hershey, PA: IGI Global. doi:10.4018/978-1-5225-0861-8.ch002

Pérez-Uribe, R. I., Torres, D. A., Jurado, S. P., & Prada, D. M. (2018). Cloud Tools for the Development of Project Management in SMEs. In R. Perez-Uribe, C. Salcedo-Perez, & D. Ocampo-Guzman (Eds.), *Handbook of Research on Intrapreneurship and Organizational Sustainability in SMEs* (pp. 95–120). Hershey, PA: IGI Global. doi:10.4018/978-1-5225-3543-0.ch005

Petrisor, I., & Cozmiuc, D. (2017). Global Supply Chain Management Organization at Siemens in the Advent of Industry 4.0. In L. Saglietto & C. Cezanne (Eds.), *Global Intermediation and Logistics Service Providers* (pp. 123–142). Hershey, PA: IGI Global. doi:10.4018/978-1-5225-2133-4.ch007

Pierce, J. M., Velliaris, D. M., & Edwards, J. (2017). A Living Case Study: A Journey Not a Destination. In N. Silton (Ed.), *Exploring the Benefits of Creativity in Education, Media, and the Arts* (pp. 158–178). Hershey, PA: IGI Global. doi:10.4018/978-1-5225-0504-4.ch008

Pipia, S., & Pipia, S. (2021). Challenges of Religious Tourism in the Conflict Region: An Example of Jerusalem. In E. Alaverdov & M. Bari (Eds.), *Global Development of Religious Tourism* (pp. 135-148). IGI Global. https://doi.org/10.4018/978-1-7998-5792-1.ch009

Poulaki, P., Kritikos, A., Vasilakis, N., & Valeri, M. (2022). The Contribution of Female Creativity to the Development of Gastronomic Tourism in Greece: The Case of the Island of Naxos in the South Aegean Region. In M. Valeri (Ed.), *New Governance and Management in Touristic Destinations* (pp. 246–258). IGI Global. https://doi.org/10.4018/978-1-6684-3889-3.ch015

Radosavljevic, M., & Andjelkovic, A. (2017). Multi-Criteria Decision Making Approach for Choosing Business Process for the Improvement: Upgrading of the Six Sigma Methodology. In J. Stanković, P. Delias, S. Marinković, & S. Rochhia (Eds.), *Tools and Techniques for Economic Decision Analysis* (pp. 225–247). Hershey, PA: IGI Global. doi:10.4018/978-1-5225-0959-2.ch011

Radovic, V. M. (2017). Corporate Sustainability and Responsibility and Disaster Risk Reduction: A Serbian Overview. In M. Camilleri (Ed.), *CSR 2.0 and the New Era of Corporate Citizenship* (pp. 147–164). Hershey, PA: IGI Global. doi:10.4018/978-1-5225-1842-6.ch008

Raghunath, K. M., Devi, S. L., & Patro, C. S. (2018). Impact of Risk Assessment Models on Risk Factors: A Holistic Outlook. In K. Strang, M. Korstanje, & N. Vajjhala (Eds.), *Research, Practices, and Innovations in Global Risk and Contingency Management* (pp. 134–153). Hershey, PA: IGI Global. doi:10.4018/978-1-5225-4754-9.ch008

Raman, A., & Goyal, D. P. (2017). Extending IMPLEMENT Framework for Enterprise Information Systems Implementation to Information System Innovation. In M. Tavana (Ed.), *Enterprise Information Systems and the Digitalization of Business Functions* (pp. 137–177). Hershey, PA: IGI Global. doi:10.4018/978-1-5225-2382-6.ch007

Rao, Y., & Zhang, Y. (2017). The Construction and Development of Academic Library Digital Special Subject Databases. In L. Ruan, Q. Zhu, & Y. Ye (Eds.), *Academic Library Development and Administration in China* (pp. 163–183). Hershey, PA: IGI Global. doi:10.4018/978-1-5225-0550-1.ch010

Ravasan, A. Z., Mohammadi, M. M., & Hamidi, H. (2018). An Investigation Into the Critical Success Factors of Implementing Information Technology Service Management Frameworks. In K. Jakobs (Ed.), *Corporate and Global Standardization Initiatives in Contemporary Society* (pp. 200–218). Hershey, PA: IGI Global. doi:10.4018/978-1-5225-5320-5.ch009

Rezaie, S., Mirabedini, S. J., & Abtahi, A. (2018). Designing a Model for Implementation of Business Intelligence in the Banking Industry. *International Journal of Enterprise Information Systems, 14*(1), 77–103. doi:10.4018/IJEIS.2018010105

Richards, V., Matthews, N., Williams, O. J., & Khan, Z. (2021). The Challenges of Accessible Tourism Information Systems for Tourists With Vision Impairment: Sensory Communications Beyond the Screen. In C. Eusébio, L. Teixeira, & M. Carneiro (Eds.), *ICT Tools and Applications for Accessible Tourism* (pp. 26–54). IGI Global. https://doi.org/10.4018/978-1-7998-6428-8.ch002

Rodrigues de Souza Neto, V., & Marques, O. (2021). Rural Tourism Fostering Welfare Through Sustainable Development: A Conceptual Approach. In A. Perinotto, V. Mayer, & J. Soares (Eds.), *Rebuilding and Restructuring the Tourism Industry: Infusion of Happiness and Quality of Life* (pp. 38–57). IGI Global. https://doi.org/10.4018/978-1-7998-7239-9.ch003

Romano, L., Grimaldi, R., & Colasuonno, F. S. (2017). Demand Management as a Success Factor in Project Portfolio Management. In L. Romano (Ed.), *Project Portfolio Management Strategies for Effective Organizational Operations* (pp. 202–219). Hershey, PA: IGI Global. doi:10.4018/978-1-5225-2151-8.ch008

Rubio-Escuderos, L., & García-Andreu, H. (2021). Competitiveness Factors of Accessible Tourism E-Travel Agencies. In C. Eusébio, L. Teixeira, & M. Carneiro (Eds.), *ICT Tools and Applications for Accessible Tourism* (pp. 196–217). IGI Global. https://doi.org/10.4018/978-1-7998-6428-8.ch009

Rucci, A. C., Porto, N., Darcy, S., & Becka, L. (2021). Smart and Accessible Cities?: Not Always – The Case for Accessible Tourism Initiatives in Buenos Aries and Sydney. In C. Eusébio, L. Teixeira, & M. Carneiro (Eds.), *ICT Tools and Applications for Accessible Tourism* (pp. 115–145). IGI Global. https://doi.org/10.4018/978-1-7998-6428-8.ch006

Ruhi, U. (2018). Towards an Interdisciplinary Socio-Technical Definition of Virtual Communities. In M. Khosrow-Pour, D.B.A. (Ed.), Encyclopedia of Information Science and Technology, Fourth Edition (pp. 4278-4295). Hershey, PA: IGI Global. doi:10.4018/978-1-5225-2255-3.ch371

Ryan, L., Catena, M., Ros, P., & Stephens, S. (2021). Designing Entrepreneurial Ecosystems to Support Resource Management in the Tourism Industry. In V. Costa, A. Moura, & M. Mira (Eds.), *Handbook of Research on Human Capital and People Management in the Tourism Industry* (pp. 265–281). IGI Global. https://doi.org/10.4018/978-1-7998-4318-4.ch013

Sabuncu, I. (2021). Understanding Tourist Perceptions and Expectations During Pandemic Through Social Media Big Data. In M. Demir, A. Dalgıç, & F. Ergen (Eds.), *Handbook of Research on the Impacts and Implications of COVID-19 on the Tourism Industry* (pp. 330–350). IGI Global. https://doi.org/10.4018/978-1-7998-8231-2.ch016

Safari, M. R., & Jiang, Q. (2018). The Theory and Practice of IT Governance Maturity and Strategies Alignment: Evidence From Banking Industry. *Journal of Global Information Management, 26*(2), 127–146. doi:10.4018/JGIM.2018040106

Sahoo, J., Pati, B., & Mohanty, B. (2017). Knowledge Management as an Academic Discipline: An Assessment. In B. Gunjal (Ed.), *Managing Knowledge and Scholarly Assets in Academic Libraries* (pp. 99–126). Hershey, PA: IGI Global. doi:10.4018/978-1-5225-1741-2.ch005

Saini, D. (2017). Relevance of Teaching Values and Ethics in Management Education. In N. Baporikar (Ed.), *Management Education for Global Leadership* (pp. 90–111). Hershey, PA: IGI Global. doi:10.4018/978-1-5225-1013-0.ch005

Sambhanthan, A. (2017). Assessing and Benchmarking Sustainability in Organisations: An Integrated Conceptual Model. *International Journal of Systems and Service-Oriented Engineering*, *7*(4), 22–43. doi:10.4018/IJSSOE.2017100102

Sambhanthan, A., & Potdar, V. (2017). A Study of the Parameters Impacting Sustainability in Information Technology Organizations. *International Journal of Knowledge-Based Organizations*, *7*(3), 27–39. doi:10.4018/IJKBO.2017070103

Sánchez-Fernández, M. D., & Manríquez, M. R. (2018). The Entrepreneurial Spirit Based on Social Values: The Digital Generation. In P. Isaias & L. Carvalho (Eds.), *User Innovation and the Entrepreneurship Phenomenon in the Digital Economy* (pp. 173–193). Hershey, PA: IGI Global. doi:10.4018/978-1-5225-2826-5.ch009

Sanchez-Ruiz, L., & Blanco, B. (2017). Process Management for SMEs: Barriers, Enablers, and Benefits. In M. Vemić (Ed.), *Optimal Management Strategies in Small and Medium Enterprises* (pp. 293–319). Hershey, PA: IGI Global. doi:10.4018/978-1-5225-1949-2.ch014

Sanz, L. F., Gómez-Pérez, J., & Castillo-Martinez, A. (2018). Analysis of the European ICT Competence Frameworks. In V. Ahuja & S. Rathore (Eds.), *Multidisciplinary Perspectives on Human Capital and Information Technology Professionals* (pp. 225–245). Hershey, PA: IGI Global. doi:10.4018/978-1-5225-5297-0.ch012

Sarvepalli, A., & Godin, J. (2017). Business Process Management in the Classroom. *Journal of Cases on Information Technology*, *19*(2), 17–28. doi:10.4018/JCIT.2017040102

Saxena, G. G., & Saxena, A. (2021). Host Community Role in Medical Tourism Development. In M. Singh & S. Kumaran (Eds.), *Growth of the Medical Tourism Industry and Its Impact on Society: Emerging Research and Opportunities* (pp. 105–127). IGI Global. https://doi.org/10.4018/978-1-7998-3427-4.ch006

Saygili, E. E., Ozturkoglu, Y., & Kocakulah, M. C. (2017). End Users' Perceptions of Critical Success Factors in ERP Applications. *International Journal of Enterprise Information Systems*, *13*(4), 58–75. doi:10.4018/IJEIS.2017100104

Saygili, E. E., & Saygili, A. T. (2017). Contemporary Issues in Enterprise Information Systems: A Critical Review of CSFs in ERP Implementations. In M. Tavana (Ed.), *Enterprise Information Systems and the Digitalization of Business Functions* (pp. 120–136). Hershey, PA: IGI Global. doi:10.4018/978-1-5225-2382-6.ch006

Schwaiger, K. M., & Zehrer, A. (2021). The COVID-19 Pandemic and Organizational Resilience in Hospitality Family Firms: A Qualitative Approach. In A. Zehrer, G. Glowka, K. Schwaiger, & V. Ranacher-Lackner (Eds.), *Resiliency Models and Addressing Future Risks for Family Firms in the Tourism Industry* (pp. 32–49). IGI Global. https://doi.org/10.4018/978-1-7998-7352-5.ch002

Scott, N., & Campos, A. C. (2022). Cognitive Science of Tourism Experiences. In R. Augusto Costa, F. Brandão, Z. Breda, & C. Costa (Eds.), Planning and Managing the Experience Economy in Tourism (pp. 1-21). IGI Global. https://doi.org/ doi:10.4018/978-1-7998-8775-1.ch001

Seidenstricker, S., & Antonino, A. (2018). Business Model Innovation-Oriented Technology Management for Emergent Technologies. In M. Khosrow-Pour, D.B.A. (Ed.), Encyclopedia of Information Science and Technology, Fourth Edition (pp. 4560-4569). Hershey, PA: IGI Global. doi:10.4018/978-1-5225-2255-3.ch396

Selvi, M. S. (2021). Changes in Tourism Sales and Marketing Post COVID-19. In M. Demir, A. Dalgıç, & F. Ergen (Eds.), *Handbook of Research on the Impacts and Implications of COVID-19 on the Tourism Industry* (pp. 437–460). IGI Global. doi:10.4018/978-1-7998-8231-2.ch021

Senaratne, S., & Gunarathne, A. D. (2017). Excellence Perspective for Management Education from a Global Accountants' Hub in Asia. In N. Baporikar (Ed.), *Management Education for Global Leadership* (pp. 158–180). Hershey, PA: IGI Global. doi:10.4018/978-1-5225-1013-0.ch008

Sensuse, D. I., & Cahyaningsih, E. (2018). Knowledge Management Models: A Summative Review. *International Journal of Information Systems in the Service Sector*, *10*(1), 71–100. doi:10.4018/IJISSS.2018010105

Seth, M., Goyal, D., & Kiran, R. (2017). Diminution of Impediments in Implementation of Supply Chain Management Information System for Enhancing its Effectiveness in Indian Automobile Industry. *Journal of Global Information Management*, *25*(3), 1–20. doi:10.4018/JGIM.2017070101

Seyal, A. H., & Rahman, M. N. (2017). Investigating Impact of Inter-Organizational Factors in Measuring ERP Systems Success: Bruneian Perspectives. In M. Tavana (Ed.), *Enterprise Information Systems and the Digitalization of Business Functions* (pp. 178–204). Hershey, PA: IGI Global. doi:10.4018/978-1-5225-2382-6.ch008

Shaqrah, A. A. (2018). Analyzing Business Intelligence Systems Based on 7s Model of McKinsey. *International Journal of Business Intelligence Research*, *9*(1), 53–63. doi:10.4018/IJBIR.2018010104

Sharma, A. J. (2017). Enhancing Sustainability through Experiential Learning in Management Education. In N. Baporikar (Ed.), *Management Education for Global Leadership* (pp. 256–274). Hershey, PA: IGI Global. doi:10.4018/978-1-5225-1013-0.ch013

Shetty, K. P. (2017). Responsible Global Leadership: Ethical Challenges in Management Education. In N. Baporikar (Ed.), *Innovation and Shifting Perspectives in Management Education* (pp. 194–223). Hershey, PA: IGI Global. doi:10.4018/978-1-5225-1019-2.ch009

Sinthupundaja, J., & Kohda, Y. (2017). Effects of Corporate Social Responsibility and Creating Shared Value on Sustainability. *International Journal of Sustainable Entrepreneurship and Corporate Social Responsibility*, *2*(1), 27–38. doi:10.4018/IJSECSR.2017010103

Škarica, I., & Hrgović, A. V. (2018). Implementation of Total Quality Management Principles in Public Health Institutes in the Republic of Croatia. *International Journal of Productivity Management and Assessment Technologies*, *6*(1), 1–16. doi:10.4018/IJPMAT.2018010101

Skokic, V. (2021). How Small Hotel Owners Practice Resilience: Longitudinal Study Among Small Family Hotels in Croatia. In A. Zehrer, G. Glowka, K. Schwaiger, & V. Ranacher-Lackner (Eds.), *Resiliency Models and Addressing Future Risks for Family Firms in the Tourism Industry* (pp. 50–73). IGI Global. doi:10.4018/978-1-7998-7352-5.ch003

Smuts, H., Kotzé, P., Van der Merwe, A., & Loock, M. (2017). Framework for Managing Shared Knowledge in an Information Systems Outsourcing Context. *International Journal of Knowledge Management*, *13*(4), 1–30. doi:10.4018/IJKM.2017100101

Sousa, M. J., Cruz, R., Dias, I., & Caracol, C. (2017). Information Management Systems in the Supply Chain. In G. Jamil, A. Soares, & C. Pessoa (Eds.), *Handbook of Research on Information Management for Effective Logistics and Supply Chains* (pp. 469–485). Hershey, PA: IGI Global. doi:10.4018/978-1-5225-0973-8.ch025

Spremic, M., Turulja, L., & Bajgoric, N. (2018). Two Approaches in Assessing Business Continuity Management Attitudes in the Organizational Context. In N. Bajgoric (Ed.), *Always-On Enterprise Information Systems for Modern Organizations* (pp. 159–183). Hershey, PA: IGI Global. doi:10.4018/978-1-5225-3704-5.ch008

Steenkamp, A. L. (2018). Some Insights in Computer Science and Information Technology. In *Examining the Changing Role of Supervision in Doctoral Research Projects: Emerging Research and Opportunities* (pp. 113–133). Hershey, PA: IGI Global. doi:10.4018/978-1-5225-2610-0.ch005

Stipanović, C., Rudan, E., & Zubović, V. (2022). Reaching the New Tourist Through Creativity: Sustainable Development Challenges in Croatian Coastal Towns. In M. Valeri (Ed.), *New Governance and Management in Touristic Destinations* (pp. 231–245). IGI Global. https://doi.org/10.4018/978-1-6684-3889-3.ch014

Tabach, A., & Croteau, A. (2017). Configurations of Information Technology Governance Practices and Business Unit Performance. *International Journal of IT/Business Alignment and Governance*, *8*(2), 1–27. doi:10.4018/IJITBAG.2017070101

Talaue, G. M., & Iqbal, T. (2017). Assessment of e-Business Mode of Selected Private Universities in the Philippines and Pakistan. *International Journal of Online Marketing*, *7*(4), 63–77. doi:10.4018/IJOM.2017100105

Tam, G. C. (2017). Project Manager Sustainability Competence. In *Managerial Strategies and Green Solutions for Project Sustainability* (pp. 178–207). Hershey, PA: IGI Global. doi:10.4018/978-1-5225-2371-0.ch008

Tambo, T. (2018). Fashion Retail Innovation: About Context, Antecedents, and Outcome in Technological Change Projects. In I. Management Association (Ed.), Fashion and Textiles: Breakthroughs in Research and Practice (pp. 233-260). Hershey, PA: IGI Global. https://doi.org/ doi:10.4018/978-1-5225-3432-7.ch010

Tantau, A. D., & Frăţilă, L. C. (2018). Information and Management System for Renewable Energy Business. In *Entrepreneurship and Business Development in the Renewable Energy Sector* (pp. 200–244). Hershey, PA: IGI Global. doi:10.4018/978-1-5225-3625-3.ch006

Teixeira, N., Pardal, P. N., & Rafael, B. G. (2018). Internationalization, Financial Performance, and Organizational Challenges: A Success Case in Portugal. In L. Carvalho (Ed.), *Handbook of Research on Entrepreneurial Ecosystems and Social Dynamics in a Globalized World* (pp. 379–423). Hershey, PA: IGI Global. doi:10.4018/978-1-5225-3525-6.ch017

Teixeira, P., Teixeira, L., Eusébio, C., Silva, S., & Teixeira, A. (2021). The Impact of ICTs on Accessible Tourism: Evidence Based on a Systematic Literature Review. In C. Eusébio, L. Teixeira, & M. Carneiro (Eds.), *ICT Tools and Applications for Accessible Tourism* (pp. 1–25). IGI Global. doi:10.4018/978-1-7998-6428-8.ch001

Trad, A., & Kalpić, D. (2018). The Business Transformation Framework, Agile Project and Change Management. In M. Khosrow-Pour, D.B.A. (Ed.), Encyclopedia of Information Science and Technology, Fourth Edition (pp. 620-635). Hershey, PA: IGI Global. https://doi.org/ doi:10.4018/978-1-5225-2255-3.ch054

Trad, A., & Kalpić, D. (2018). The Business Transformation and Enterprise Architecture Framework: The Financial Engineering E-Risk Management and E-Law Integration. In B. Sergi, F. Fidanoski, M. Ziolo, & V. Naumovski (Eds.), *Regaining Global Stability After the Financial Crisis* (pp. 46–65). Hershey, PA: IGI Global. doi:10.4018/978-1-5225-4026-7.ch003

Trengereid, V. (2022). Conditions of Network Engagement: The Quest for a Common Good. In R. Augusto Costa, F. Brandão, Z. Breda, & C. Costa (Eds.), *Planning and Managing the Experience Economy in Tourism* (pp. 69-84). IGI Global. https://doi.org/10.4018/978-1-7998-8775-1.ch004

Turulja, L., & Bajgoric, N. (2018). Business Continuity and Information Systems: A Systematic Literature Review. In N. Bajgoric (Ed.), *Always-On Enterprise Information Systems for Modern Organizations* (pp. 60–87). Hershey, PA: IGI Global. doi:10.4018/978-1-5225-3704-5.ch004

Vargas-Hernández, J. G. (2017). Professional Integrity in Business Management Education. In N. Baporikar (Ed.), *Management Education for Global Leadership* (pp. 70–89). Hershey, PA: IGI Global. doi:10.4018/978-1-5225-1013-0.ch004

Varnacı Uzun, F. (2021). The Destination Preferences of Foreign Tourists During the COVID-19 Pandemic and Attitudes Towards: Marmaris, Turkey. In M. Demir, A. Dalgıç, & F. Ergen (Eds.), *Handbook of Research on the Impacts and Implications of COVID-19 on the Tourism Industry* (pp. 285–306). IGI Global. https://doi.org/10.4018/978-1-7998-8231-2.ch014

Vasista, T. G., & AlAbdullatif, A. M. (2017). Role of Electronic Customer Relationship Management in Demand Chain Management: A Predictive Analytic Approach. *International Journal of Information Systems and Supply Chain Management, 10*(1), 53–67. doi:10.4018/IJISSCM.2017010104

Vieru, D., & Bourdeau, S. (2017). Survival in the Digital Era: A Digital Competence-Based Multi-Case Study in the Canadian SME Clothing Industry. *International Journal of Social and Organizational Dynamics in IT, 6*(1), 17–34. doi:10.4018/IJSODIT.2017010102

Vijayan, G., & Kamarulzaman, N. H. (2017). An Introduction to Sustainable Supply Chain Management and Business Implications. In M. Khan, M. Hussain, & M. Ajmal (Eds.), *Green Supply Chain Management for Sustainable Business Practice* (pp. 27–50). Hershey, PA: IGI Global. doi:10.4018/978-1-5225-0635-5.ch002

Vlachvei, A., & Notta, O. (2017). Firm Competitiveness: Theories, Evidence, and Measurement. In A. Vlachvei, O. Notta, K. Karantininis, & N. Tsounis (Eds.), *Factors Affecting Firm Competitiveness and Performance in the Modern Business World* (pp. 1–42). Hershey, PA: IGI Global. doi:10.4018/978-1-5225-0843-4.ch001

Wang, C., Schofield, M., Li, X., & Ou, X. (2017). Do Chinese Students in Public and Private Higher Education Institutes Perform at Different Level in One of the Leadership Skills: Critical Thinking?: An Exploratory Comparison. In V. Wang (Ed.), *Encyclopedia of Strategic Leadership and Management* (pp. 160–181). Hershey, PA: IGI Global. doi:10.4018/978-1-5225-1049-9.ch013

Wang, J. (2017). Multi-Agent based Production Management Decision System Modelling for the Textile Enterprise. *Journal of Global Information Management, 25*(4), 1–15. doi:10.4018/JGIM.2017100101

Wiedemann, A., & Gewald, H. (2017). Examining Cross-Domain Alignment: The Correlation of Business Strategy, IT Management, and IT Business Value. *International Journal of IT/Business Alignment and Governance, 8*(1), 17–31. doi:10.4018/IJITBAG.2017010102

Wolf, R., & Thiel, M. (2018). Advancing Global Business Ethics in China: Reducing Poverty Through Human and Social Welfare. In S. Hipsher (Ed.), *Examining the Private Sector's Role in Wealth Creation and Poverty Reduction* (pp. 67–84). Hershey, PA: IGI Global. doi:10.4018/978-1-5225-3117-3.ch004

Yablonsky, S. (2018). Innovation Platforms: Data and Analytics Platforms. In *Multi-Sided Platforms (MSPs) and Sharing Strategies in the Digital Economy: Emerging Research and Opportunities* (pp. 72–95). Hershey, PA: IGI Global. doi:10.4018/978-1-5225-5457-8.ch003

Yaşar, B. (2021). The Impact of COVID-19 on Volatility of Tourism Stocks: Evidence From BIST Tourism Index. In M. Demir, A. Dalgıç, & F. Ergen (Eds.), *Handbook of Research on the Impacts and Implications of COVID-19 on the Tourism Industry* (pp. 23–44). IGI Global. https://doi.org/10.4018/978-1-7998-8231-2.ch002

Yusoff, A., Ahmad, N. H., & Halim, H. A. (2017). Agropreneurship among Gen Y in Malaysia: The Role of Academic Institutions. In N. Ahmad, T. Ramayah, H. Halim, & S. Rahman (Eds.), *Handbook of Research on Small and Medium Enterprises in Developing Countries* (pp. 23–47). Hershey, PA: IGI Global. doi:10.4018/978-1-5225-2165-5.ch002

Zacher, D., & Pechlaner, H. (2021). Resilience as an Opportunity Approach: Challenges and Perspectives for Private Sector Participation on a Community Level. In A. Zehrer, G. Glowka, K. Schwaiger, & V. Ranacher-Lackner (Eds.), *Resiliency Models and Addressing Future Risks for Family Firms in the Tourism Industry* (pp. 75–102). IGI Global. https://doi.org/10.4018/978-1-7998-7352-5.ch004

Zanin, F., Comuzzi, E., & Costantini, A. (2018). The Effect of Business Strategy and Stock Market Listing on the Use of Risk Assessment Tools. In *Management Control Systems in Complex Settings: Emerging Research and Opportunities* (pp. 145–168). Hershey, PA: IGI Global. doi:10.4018/978-1-5225-3987-2.ch007

Zgheib, P. W. (2017). Corporate Innovation and Intrapreneurship in the Middle East. In P. Zgheib (Ed.), *Entrepreneurship and Business Innovation in the Middle East* (pp. 37–56). Hershey, PA: IGI Global. doi:10.4018/978-1-5225-2066-5.ch003

About the Contributors

Fernando Ortiz-Rodriguez is a full professor at Tamaulipas Autonomous University, a Member of National Research Council level C, and director of the Social Research Centre at UAT, Mexico. He is a member of the Information Technology research group and part of the Knowledge Graph and Semantic Web Community. He had published, Journal Articles, Book Chapters, and as a book editor from reputed editorials as Springer and Elsevier. Also, he had presented many conference articles in Europe, the United States, and Mexico. Main chair and organizer of KGSWC multi-series conference.

Sanju Mishra Tiwari has worked as a Research Associate for a sponsored research project "Intelligent Real time Situation Awareness and Decision Support System for Indian Defence" funded by DRDO in Department of Computer Applications, National Institute of Technology, Kurukshetra. Her current research interests include Knowledge Based Systems, Ontologism, Intelligent Decision Support, Semantic/ Intelligent Query Search Engines, Intrusion Detection, Swarm Intelligence. She has to-date published 19 research papers one 1 book chapter with international and national publishers. She has 10 years experience in teaching field and 2 years for research field for developing ontologies. She is the member of IEEE. She is working as an organizing committee member of MIR Labs, USA.

Luis Hernandez-Gonzalez has a Bachelor's degree in accounting, MBA in Administration, and PhD in business. Experience includes financial analysis, company setup, lean accounting, and lean manufacturing. More than 20 years of experience in manufacturing and finance in medical devices, recreational goods, and appliances. Expertise includes non-profit organization management and binational representation. Have supported company start-ups and investor relations.

Agustín Tiburcio-Sanchez is an Electronic Systems Engineer from the Tecnológico de Monterrey campus Toluca and has a Master's degree Studies in Artificial Intelligence from the Universidad Veracruzana. He has the degree of Master Black Belt in process improvement tools the Six Sigma methodology. He has more than 25 years of experience in the areas of systems, technology and the maquiladora industry. He has been in charge of the IT department at Gentel and the systems manager at Falcomex, in the state of Coahuila, systems leader at Siemens VDO in Monterrey, senior systems manager at Alcom Electrónica de México at Reynosa, Tamaulipas, director of systems at ARC Automotriz, director of business development at Transtelco for the Northwest area and commercial manager national wise for Mexico Industry magan¡zine He has collaborated closely with Index Reynosa, since 2007, in various positions. In 2010-11, former president of the Information Technology Committee. During his tenure, among some other activities, he organized different national and international technology forums. He has participated in municipal and state technological promotion citizen councils. He is director of the systems department of Global Industrial Manufacturing as well as director of Convertec, an information technology consultancy, which has advised companies such as ZafiroSoft, Link Technologies, SMTELCOM, Creatividad y Enlaces and Automotive News México. As of October 2021, he serves as director of the Information Technology committee at the national wise for the national council of the index maquiladora and manufacturing industry, leading forceful actions such as linking the index with ITESM and UAT, as well as the collaboration agreement with CANIETI and the planning, direction, and execution of the first National Cybersecurity Forum of index "Technology has always been my passion and I have practically dedicated my life to using it to improve processes and the lives of others in companies and in society"

* * *

Shahaas A is a student pursuing his Bachelors in Mechanical Engineering at Sahyadri College of Engineering & Management. He has three SCOPUS publications in CPS and robotics.

Daniel Avila-Guzman is a free-time Research Professor. Autonomous University of Tamaulipas. Reynosa-Rodhe Multidisciplinary Academic Unit. Reynosa, Tamaulipas, Mexico

G. Balamurugan is working as an Assistant Professor, in the Department of Computing Technologies, SRM Institute of Science and Technology Kattankulathur Campus Chengalpattu-603203.

C.P. Dheeshith is a Graduate Engineer Trainee at ATKINS. He has three SCOPUS publications in CPS and robotics.

Edgar Garcia has a Master degree on productivity & quality administration June, 2023. Bachelor degree on Mechatronic engineering September,2015.

Rathishchandra R Gatti is currently working as a Professor & head of the Department of Mechanical Engineering and Robotics and Automation. Prior to this, he has worked as IP Analyst at Curtin University, Perth Australia, NPI program manager at General Electric for Industrial Switchgear Products, Production Engineer at SVE and Rapsri and Design Engineer at Meritor Automotive. He pursued his bachelors in Mechanical Engineering at NITK Suratkal, MBA from IGNOU, PGDGSM from ICFAI and PhD from Curtin University,Perth Australia. He has 4 Patents (1US, 1 Canadian,1 Mexican,1 Indian), 40+ peer-reviewed publications and is the editor of 7 research books and one SCOPUS journal. He has over 20 years of R&D experience in the Mechanical and Mechatronics domains. His research interests include Energy Harvesting, Robotics, Self-Powered CPS, and AI in CPS.

Shyam J S is an Assistant professor, Dept. Of CSE, GITAM University Bengaluru.

Abhijith K is working at Cognizant Technology Solutions India Pvt Ltd. He has three SCOPUS publications in CPS and robotics.

José Melchor has a PhD in Business Information Systems from the Polytechnic University of Madrid (Spain) and expert in Information Technologies and Systems. Leader of the research group Information Technologies and Strategy (UAT-CA-132), for which he has published 22 papers in high-impact journals (Scopus and JCR) both in Mexico and internationall: Journal of Computer Information Systems, Journal of Information Systems and Technology Management, Innovar Journal, Cuadernos de Administración, Management Studies, Nova Scientia, Accounting and Administration, among others. He is the co-author of 20 book chapters on information technologies and systems. He wrote the book Information Systems Project Planning, co-authored a leading-edge statistical book such as Multivariate Analysis Techniques. Applied to the social sciences, co-author in 6 books with a business administration approach, and coordinator of 5 books. He has supervised 73 theses between graduate, master, and doctorate.

Olegario Mendez Cabrera is a full-time Research Professor. Autonomous University of Tamaulipas. Reynosa-Rodhe Multidisciplinary Academic Unit. Reynosa, Tamaulipas, Mexico. Email: olmendez@docentes.uat.edu.mx

Samantha Méndez-Sánchez has more than 10 years of experience, working in both the public and private sectors. My professional development is focused on the area of institutional and governmental relations, in the management of corporate projects in various matters, and in the coordination of committees and national working groups projects; as well as in the organization of business events at the national level (forums, webinars and seminars), both face-to-face and virtual. Currently, I coordinate part of the national network of Committees and Working Groups of the National Council of the Maquiladora and Export Manufacturing Industry (index), a business organization that represents more than 1,300 global companies that are beneficiaries of the IMMEX program.

Francisco Isaí Morales-Sáenz is a professor-researcher at the Faculty of Commerce and Administration Victoria. His research line is Information Technologies and Business Strategy. Email: fmsaenz@uat.edu.mx

Rithin Nambiar is working at Bosch Automotive Electronics India Pvt Ltd. He has three SCOPUS publications in CPS and robotics.

Susuna Orallo is the General Director Irium Mexico General Counsel at CAMESCOM. (Spanish Chamber of Commerce in Mexico). Advisory Council of the Spanish Ambassador in Mexico. President of the Human Resources Commission CAMESCOM (Spanish Chamber of Commerce in Mexico). President of the CAMESCOM Government Relations and Networking Sub-Commission. NATIONAL INDEX Technology Commission, Cybersecurity Subcommittee Directorate. NuBE Foundation Board of Directors (Nutrition, Well-being and Education for children and adolescents).

Disha Patil is a graduate student at Northeastern University, Boston pursuing a Master of Science in Information Systems.

Alexei Pinal has a bachelor degree in TeleCommunications and Electronics Engineer by IPN (Instituto Politécnico Nacional) and has senior management studies at IPADE in Mexico Alexei worked in global automation companies providing electrical and process automation solutions for the Chemical, M&M, Pharmaceutical, O&G, Manufacturing and Food industries; performing different roles such as Engineering development, Project Management, BDM and Sales in projects in India, Holland, Norway, Germany, USA and Mexico.

Miguel A. Sahagun is an Assistant Professor of Marketing at High Point University (HPU), Dr. Sahagun obtained his Ph.D. from The University of Texas-Pan American (UTPA), his Master of Business Administration from the University of Texas-Brownsville (UTB), and his Bachelor's degree in Industrial Engineering from the Instituto Tecnologico y de Estudios Superiores de Occidente (ITESO). Dr. Sahagun has 25 years of corporate and academic experience. He is an active member of AMA, SMA, and ISoF. Dr. Sahagun has published and presented his research at various scientific journals, book chapters, and international conferences in Peru, Mexico, Romania, Ecuador, Cuba, Canada, China, Chile, Portugal, Morocco, Greece, and the United States of America.

Chandra Singh is currently working as Assistant Professor at Sahyadri college of Engineering & Management. He has obtained M.Tech in Digital Electronics & Communication from NMAM Institute of Technology Nitte. He has obtained B.E in Electronics & Communication Engineering at Srinivas School of Engineering, Mukka,Mangaluru. He has Co-Authored one textbook entitled "Survivability Techniques in Optical Network" published by Studuim Press. He has published more 15 papers in Scopus & Web of Science Journals.

Amit Kumar Tyagi (GATE, NPwD-JRF, UGC-NET, and ICAR-NET) received his Ph.D degree in 2018 from Pondicherry Central University, Puducherry, India, in area of "Vehicular Ad-hoc Networks". His research interests include Formal Language Theory, Smart and Secure Computing, Privacy (including Genomic Privacy), Machine Learning with Big data, Blockchain Technology, Cyber Physical System etc. He has completed his M.Tech degree from the Pondicherry Central University, Puducherry, India. He joined the Lord Krishna College of Engineering, Ghaziabad (LKCE) for the periods of 2009-2010, and 2012-2013. With more than 08 (Eight) years of teaching and research experience across India, currently he is working as an Assistant Professor in Vellore Institute of Technology, Chennai Campus, Chennai 600127, Tamilnadu, India. Additionally, He is also the recipient/ awarded of the GATE and NPwD-JRF fellowship in 2009, 2016 and 2013. He has been published one major book titled "Know Your Technical (IT) Skills". Also He is a member of various Computer/ Research Communities like IEEE, ISOC, CSI, ISTE, DataScience, MIRLab, etc.

Index

T

W

Z

www.ingramcontent.com/pod-product-compliance
Lightning Source LLC
LaVergne TN
LVHW081331300926
849161LV00008B/1944
* 9 7 8 1 6 6 8 4 8 0 8 8 5 *